Alexandra Duarte
Nathalie Zamora
Carlos Jacomino

THE NEW ERA OF CONSTRUCTION

Alexandra Duarte
Nathalie Zamora
Carlos Jacomino

THE NEW ERA OF CONSTRUCTION

Game-changing materials

ScienciaScripts

Imprint

Cover image: www.ingimage.com

This book is a translation from the original published under ISBN 978-613-9-43875-4.

Publisher:
Sciencia Scripts
is a trademark of
Dodo Books Indian Ocean Ltd. and OmniScriptum S.R.L publishing group

120 High Road, East Finchley, London, N2 9ED, United Kingdom
Str. Armeneasca 28/1, office 1, Chisinau MD-2012, Republic of Moldova, Europe
Managing Directors: Ieva Konstantinova, Victoria Ursu
info@omniscriptum.com

Printed at: see last page
ISBN: 978-620-8-38800-3

Index

For the reader

Welcome to a journey into the future of construction, where technological innovation and sustainability meet to redefine the materials and methods that will shape the cities of tomorrow. This book is a guide to the emerging technologies that are revolutionising the construction industry, from self-healing concrete to nanotechnology and 3D printing. Each chapter explores a key aspect of these innovations, giving you the fundamentals, practical applications, benefits and challenges facing each technology.

The organisation of this book will allow you to delve into each technology in a structured way: we start with the key concepts, follow with applied examples and conclude with an analysis of their potential impact. This structure is designed to help you not only understand the materials, but also to visualise how they can be integrated into the design and construction of sustainable and durable projects.

As you read on, you will find that this book is divided into chapters dedicated to each disruptive technology. We start with self-healing concrete, a material that, using principles of chemistry, biology and physics, can heal cracks autonomously, prolonging the life of structures. Next, we dive into nanotechnology, which enables the design of incredibly strong and adaptable materials. Finally, 3D printing and BIM methodology are presented as fundamental tools for the efficiency, customisation and digitisation of construction projects.

As you read, you will see how these technologies face not only technical challenges, but also the need for a change of mindset in the industry. By the end of the book, we hope you will not only have a comprehensive overview of the current state of these innovations, but also an understanding of how they can be integrated into future projects in an effective and sustainable way.

CAPÍTULO 1

Hormigón autorreparable

1.1 Introduction

Concrete is one of the most widely used building materials in the world, accounting for approximately 70% of all building materials in terms of volume. Its popularity is due to its strength, durability and versatility, making it the preferred choice for a wide range of structures, from residential buildings to critical infrastructure such as bridges and tunnels. However, despite its many advantages, concrete's durability is compromised by the appearance of cracks and damage over time, which can result in costly repairs and a reduction in the service life of the material.

Cracks in concrete can be caused by multiple factors, including temperature changes, ground settlement, structural loads and chemical aggressions. These cracks not only affect the aesthetics of buildings, but can also allow water infiltration and corrosion of steel reinforcement, leading to accelerated deterioration of the structure. According to a study by the American Society of Civil Engineers (2017), the cost of repairing deteriorated infrastructure in the United States is estimated at hundreds of billions of dollars annually. This context has led to the search for innovative solutions that can effectively address these problems.

In this regard, the development of self-healing concrete has emerged as a promising solution to mitigate problems related to cracks in concrete. This type of concrete has the ability to automatically heal the cracks that occur, which not only prolongs the service life of buildings, but also significantly reduces maintenance and repair costs. Self-repairing concrete represents a significant advance in building materials technology, combining principles of civil engineering, chemistry and biology to create a more resilient material.

This chapter explores in depth the principles behind self-repairing concrete, the different types that exist, its advantages and applications, as well as the challenges it faces in its implementation. In addition, the potential of this innovation to transform the construction industry will be addressed, highlighting its relevance in the current context of sustainability and resource efficiency.

Figure 1

Self-repairing concrete photo

Source: Technical University of Delft

1.2 Principles of Self-Repairing Concrete

Self-repairing concrete is an innovation in materials engineering that seeks to improve the durability and sustainability of concrete structures. The fundamental principles underpinning this technology are detailed below, each amplified with relevant examples and considerations.

1.2.1 Self-repair chemistry

Self-repair chemistry is based on the ability of concrete to react chemically with components in its environment to seal cracks and fissures. This principle focuses on the incorporation of admixtures that facilitate specific chemical reactions when in contact with water and other elements.

When a crack forms, curing agents, which may include silicates and carbonate compounds, are activated. For example, the use of additives such as sodium silicate can result in the formation of a gel that fills the crack. This gel not only seals the crack, but can also react with CO_2 present in the environment to form calcium carbonate, which reinforces the original structure (Böngen et al., 2018). Self-repairing chemistry enables the development of concretes that are not only able to seal cracks, but can also improve their compressive and tensile strength after the curing process.

In addition, research in the field of chemical admixtures has led to the development of concretes that can cure in different environmental conditions. For example, some admixtures are more effective in humid climates, where the presence of water activates the curing reactions, while others can be designed to work in drier conditions. This customised approach to self-healing concrete chemistry allows for its application in a variety of environments, from coastal construction to desert structures.

The ability of these concretes to adapt to different environments not only improves repair efficiency, but also adds to the sustainability of the material. By reducing the need for frequent repairs, resource consumption and waste generation is reduced, which is essential in today's context of environmental sustainability.

1.2.2 Biology and Microbiology

The use of biological principles in self-repairing concrete represents one of the most promising innovations in the construction sector. This principle is based on the incorporation of micro-organisms that can be activated under specific conditions, such as the presence of water and nutrients, to repair damage to concrete.

Bacteria, such as Bacillus subtilis, are a notable example. These bacteria can survive in adverse conditions and are able to produce calcium carbonate through metabolic processes. When cracks form, water penetrates the concrete, activating bacteria that begin to metabolise available nutrients and produce calcium carbonate, which is deposited in the crack (Jonkers, 2011). This process not only seals the crack, but can also strengthen the surrounding concrete structure.

Biology applied to self-repairing concrete opens up new avenues for sustainability in construction. On the one hand, the use of micro-organisms reduces reliance on traditional repair methods, which are often costly and resource-intensive. On the other hand, this approach allows for a longer life cycle for concrete structures, which is essential in a world where longevity of materials is increasingly important.

In addition, research in this area has led to the development of "smart concretes" that not only repair themselves, but can also adapt to changes in their environment. For example, methods are being explored to modify bacteria to respond to different environmental stimuli, which could further optimise their ability to self-repair.

1.2.3 Physics of Materials

The physics of materials is fundamental to the development of self-healing concrete, as it refers to how the physical properties of concrete, such as its porosity, density and strength, affect its ability to self-heal. This principle focuses on the design and formulation of concrete to maximise its effectiveness in self-healing.

The porosity of the concrete is a critical factor. Concrete with an adequate porous structure allows curing agents, whether chemical or biological, to reach the cracks effectively. If the concrete is too dense or compact, curing agents may not have access to the cracks, limiting their ability to make repairs (Mechtcherine et al., 2016). Therefore, the design of self-healing concrete must balance strength and porosity to facilitate this process.

In addition, research into the physics of materials has led to the creation of concretes that respond to changing environmental conditions. For example, concretes are being developed that can expand or contract in response to temperature changes, which can help seal cracks more effectively. This principle focuses on adjusting the composition of the concrete by incorporating admixtures that allow for these dynamic properties.

The physics of materials also extends to understanding how external forces, such as structural loads and vibrations, affect the integrity of concrete. By understanding these interactions, engineers can design concretes that are not only resistant to cracking, but are also capable of repairing themselves effectively when damage occurs.

Materials Engineering

Materials engineering is a key principle in the development of self-healing concrete, which refers to the selection and combination of materials that improve the properties of concrete. This approach is essential to create a material that is not only strong and durable, but also capable of effective self-repair.

Research in this field has led to the identification of admixtures and compounds that can improve the cohesion and strength of concrete. For example, the incorporation of synthetic or natural fibres can increase tensile strength and reduce crack formation. This materials engineering principle enables the design of concretes that are more resilient to adverse environmental conditions (Huang et al., 2015).

In addition, the use of new materials, such as nanomaterials and polymers, has shown great potential in improving the properties of self-healing concrete. Nanomaterials, in particular, can improve the corrosion resistance and durability of concrete, which is essential for structures exposed to aggressive environments, such as coastlines and industrial areas.

Materials engineering also involves the creation of concretes with specific properties for particular applications. For example, lightweight concretes can be developed for high-rise building applications, which require a balance between strength and weight. Or, alternatively, high-strength concretes for structures supporting heavy loads. This customised approach to the formulation of self-repairing concrete allows its use in a wide range of applications.

1.2.4 Monitoring and Response to Environmental Conditions

The principle of monitoring and response refers to the ability of self-healing concrete to detect environmental conditions and activate repair mechanisms. This approach takes advantage of advanced technologies, such as sensors and monitoring systems, which enable a proactive response to deterioration.

Sensors embedded in the concrete structure can continuously monitor the health of the material, detecting the appearance of cracks or changes in environmental conditions. When a crack is detected, the system can automatically activate curing agents, ensuring a fast and efficient repair (Wang et al., 2016). This principle not only improves the effectiveness of self-healing concrete, but also enables the prevention of further damage that could compromise structural integrity.

The implementation of monitoring technologies also has significant implications for infrastructure maintenance. Rather than relying on periodic inspections, continuous monitoring allows engineers and infrastructure managers to make informed decisions about the timing and type of maintenance required. This can result in more efficient use of resources and reduced costs associated with repair and maintenance.

In addition, the principle of monitoring and response can be integrated with artificial intelligence systems that analyse data to predict when and where cracks are most likely to occur. This proactive approach allows engineers to address problems before they become significant failures, improving the safety and durability of structures.

1.2.5 Sustainability and Resource Efficiency

Sustainability is a fundamental principle in the development of self-repairing concrete. In a context where construction and infrastructure contribute significantly to the global carbon footprint, self-repairing concrete offers a solution that can reduce the environmental impact.

By extending the service life of structures and reducing the need for frequent repairs, self-healing concrete contributes to a more efficient use of material resources. This not only reduces the consumption of new materials, but also reduces waste generation, which is essential for sustainable construction (Kakooei et al., 2017). In addition, this type of concrete can incorporate recycled materials in its formulation, which further contributes to the sustainability of the construction process.

The ability of self-healing concrete to maintain its structural integrity over time also means less disruption to infrastructure use. This is particularly important in urban areas, where repairs can cause congestion and affect residents' quality of life. By minimising these disruptions, self-repairing concrete not only improves environmental sustainability, but also supports social well-being.

In short, self-healing concrete is based on interdisciplinary principles that integrate chemistry, biology, physics and materials engineering to offer an innovative and sustainable solution to the challenges of modern construction. As these principles are developed and optimised, self-healing concrete is expected to transform the construction industry, providing more durable and sustainable structures.

1.3 Types of Self-Repairing Concrete

Self-repairing concrete is a significant advance in materials engineering, designed to improve the durability and sustainability of concrete structures. There are several types of self-repairing concrete, each with specific characteristics and repair mechanisms. The most prominent types, their operating principles and applications are described below.

1.3.1 Self-Repairing Concrete Based on Microorganisms

Self-repairing concrete based on micro-organisms is an innovative solution in the field of construction that aims to improve the durability and sustainability of concrete structures. This type of concrete incorporates microorganisms that, when in contact with water and nutrients, can precipitate calcium carbonate, which seals cracks and crevices that form over time. This process not only prolongs the life of the concrete, but also reduces the need for costly repairs and the use of additional materials.

The principle behind self-repairing concrete is based on bio-construction. Microorganisms, such as certain species of bacteria, are encapsulated in the concrete during mixing. When cracks occur, water penetrates the concrete, activating the micro-organisms. These microorganisms metabolise the nutrients present and, as a result, produce calcium carbonate, which is deposited in the cracks, effectively sealing them (Jonkers, 2011).

The advantages of self-repairing concrete include its sustainability, durability and cost reduction. By reducing the need for repairs and the use of new materials, self-repairing concrete contributes to more sustainable construction practices. In addition, it can significantly extend the service life of structures, which is especially beneficial in harsh environments where traditional concrete could deteriorate rapidly. Although the initial investment may be higher, long-term costs are reduced due to fewer repairs and maintenance.

However, self-healing concrete based on micro-organisms faces several challenges. One of the main ones is the appropriate selection of micro-organisms that are effective under various environmental conditions. Research on the long-term viability of these micro-organisms in concrete is still limited. It is also crucial to consider the interaction of these microorganisms with other concrete admixtures and components, which may affect their overall performance (Kakooei et al., 2017).

In conclusion, self-healing concrete based on micro-organisms represents a significant advance in civil engineering and sustainable construction. Its ability to automatically repair cracks not only improves the durability of structures, but also promotes a greener approach to construction. As research progresses and current challenges are overcome, it is likely that this type of concrete will become a standard option in the construction industry, contributing to a more sustainable and efficient future.

1.3.2 Self-Repairing Concrete with Microcapsules

Self-repairing concrete with microcapsules is an emerging technology in the field of civil engineering that seeks to address the limitations of conventional concrete, especially in terms of durability and maintenance. This type of concrete incorporates microcapsules containing curing agents, such as resins or chemicals, designed to release their contents when cracks occur in the structure. When these agents are released, a curing process is initiated to seal the cracks, restoring the integrity of the concrete and prolonging its service life.

The functioning of self-repairing concrete with microcapsules is based on the principle of encapsulation. During the manufacture of the concrete, microcapsules are added, which can be made of different materials such as polymers or ceramics. When the concrete cracks, the microcapsules break, releasing their contents into the crack. This content reacts with water and other concrete components, forming a gel or solid material that seals the crack and prevents the penetration of external agents, such as water and contaminants (Wang et al., 2016).

This type of concrete offers several significant advantages. Firstly, it improves the durability of the material and thus the service life of structures. This is especially important in environments where concrete is exposed to adverse conditions, such as freeze-thaw cycles or exposure to chemicals. In addition, the implementation of self-healing concrete with microcapsules can significantly reduce long-term maintenance costs by reducing the need for frequent and costly repairs.

However, self-repairing concrete with microcapsules also presents challenges. One of the main ones is the need to research and develop microcapsules that are effective and economical on a large scale. In addition, consideration needs to be given to how the inclusion of microcapsules affects the mechanical properties of the concrete, such as its strength and durability. Compatibility of curing agents with other concrete admixtures is also a critical aspect that needs to be studied to ensure optimal performance (Mechtcherine et al., 2016).

In conclusion, microcapsule self-healing concrete represents a major breakthrough in the search for more durable and sustainable building materials. Its ability to automatically seal cracks by releasing curing agents may revolutionise the way concrete structures are designed and maintained. As research continues and current challenges are overcome, this technology is likely to become increasingly integrated into civil engineering practice, contributing to a more resilient and efficient future in construction.

1.3.3 Self-Repairing Concrete with Expandable Polymers

Self-repairing concrete with expanding polymers is an innovation that seeks to improve the durability and resistance of concrete to cracks and fissures. This type of concrete incorporates polymer admixtures that, when in contact with water or when activated by environmental changes, expand and automatically fill the cracks that form in the structure. This self-repairing capability not only prolongs the life of the concrete, but also significantly reduces the need for manual repairs and the consumption of additional resources.

The working mechanism of self-healing concrete with expanding polymers is based on the reaction of these polymers to water or temperature changes. During the manufacture of the concrete, microcapsules or polymer particles are integrated which, when water is detected, expand. This expansion allows the polymer to fill the cracks, forming an effective seal that prevents the infiltration of moisture and other damaging agents that can compromise the structural integrity of the concrete (Le et al., 2012).

Among the advantages of using self-healing concrete with expanding polymers is its ability to improve the strength and durability of the material. The expansion of the polymers not only seals cracks, but can also help to reinforce the affected areas, resulting in a longer service life of the structures. In addition, by reducing the need for frequent repairs, long-term maintenance costs are minimised, making this technology a viable and cost-effective option for construction projects.

However, there are challenges associated with the use of expandable polymers in concrete. One of the main ones is ensuring the compatibility of these polymers with the cement and other admixtures used in the mix. Research on the durability and behaviour of these polymers over time is also essential, as consistent performance under various environmental conditions is required (Zhang et al., 2016). In addition, a detailed analysis of how the inclusion of polymers affects the mechanical properties of concrete, such as its compressive and tensile strength, is needed.

In conclusion, self-healing concrete with expanding polymers represents a significant advance in construction technology, offering an effective solution to improve concrete durability and reduce maintenance costs. Its ability to automatically seal cracks and improve structural strength may revolutionise the way infrastructure is designed and maintained. As research progresses and existing challenges are addressed, this type of concrete has the potential to become a standard option in the construction industry, contributing to a more sustainable and efficient future.

1.3.4 Self-Repairing Concrete with Chemical Additives

Self-repairing concrete with chemical admixtures is an innovation that seeks to improve the durability and longevity of concrete structures by incorporating chemicals that react to the formation of cracks. These admixtures are designed to activate when a crack is detected, allowing a gel or solid material to form and seal the crack, thereby restoring the integrity of the concrete.

The operation of this type of concrete is based on the chemistry of the admixtures. When these compounds are mixed with the concrete, they are prepared to react when water enters the cracks. For example, some chemical admixtures can release sealing agents that, when in contact with water, expand and form a material that seals the crack. This reaction can be instantaneous or progressive, depending on the formulation of the additive (Ramakrishnan et al., 2009).

The advantages of using self-healing concrete with chemical admixtures are notable. Firstly, this approach can significantly extend the service life of structures, minimising deterioration and the need for repairs. This is especially valuable in environments where concrete is exposed to adverse conditions, such as moisture and corrosion. In addition, the implementation of chemical admixtures can be more cost-effective in the long term, as it reduces maintenance costs and increases the efficiency of resources used in construction.

Despite the benefits, there are also challenges associated with the use of chemical admixtures. One of the main ones is the need to investigate the compatibility of these admixtures with the concrete components, as well as their long-term effectiveness. It is also essential to evaluate how the inclusion of these admixtures affects the mechanical properties of concrete, such as its compressive strength and durability under extreme conditions (Mechtcherine et al., 2016).

In terms of applications, self-healing concrete with chemical admixtures is suitable for a variety of structures, from buildings to bridges and pavements. Its ability to automatically repair cracks can be particularly beneficial in critical infrastructure where safety and durability are paramount. As research progresses in this area, it is expected that more effective and sustainable formulations will be developed that maximise the performance of self-repairing concrete.

1.3.5 Self-Repairing Concrete with Nanomaterials

Self-repairing concrete with nanomaterials is an innovative technology that seeks to improve the mechanical properties and durability of concrete by incorporating nanoparticles. These nano-scale materials, such as graphene oxide, carbon nanotubes and silica nanoparticles, provide unique characteristics that can optimise the performance of concrete, facilitating its ability to self-repair.

Nanomaterials can improve the cohesion and strength of concrete, as well as its impermeability. For example, silica nanoparticles can fill pores in the concrete microstructure, which reduces permeability and thus the ingress of water and aggressive agents that can cause cracks. When cracks occur, the presence of these nanomaterials can facilitate the activation of self-repair processes, as they create an environment conducive for curing agents, whether chemical or biological, to work more effectively (Zhang et al., 2016).

One of the most significant advantages of self-repairing concrete with nanomaterials is their ability to improve the durability of structures. The use of these materials can result in concrete that is more resistant to corrosion, freeze-thaw cycles and other adverse environmental factors. This is particularly relevant in infrastructure exposed to extreme conditions, such as bridges and tunnels, where material integrity is critical (Al-Tabbaa & Azzam, 2015).

However, the incorporation of nanomaterials also presents challenges. One of the main ones is the variability in the behaviour of nanomaterials depending on their origin and manufacturing process. This can affect the consistency of concrete performance and complicate large-scale production. In addition, the interaction between nanomaterials and concrete components must be carefully studied to avoid undesirable effects that may compromise the strength and durability of the material (Kaur & Singh, 2023).

Research on self-repairing concrete with nanomaterials is still in its early stages, but preliminary results are promising. As new formulations are developed and optimised, it is likely that these technologies will become more widely deployed in the construction industry, offering effective solutions to improve the sustainability and resilience of infrastructure.

1.3.6 Self-Repairing Concrete with Monitoring Systems

Self-repairing concrete with monitoring systems is an emerging trend in civil engineering that combines the self-repairing capacity of concrete with advanced monitoring technologies to ensure the structural integrity of buildings. This approach not only allows early identification of cracks, but also optimises the automatic repair process, improving the safety and durability of structures.

The implementation of monitoring systems in self-repairing concrete involves the use of integrated sensors that can detect changes in material conditions, such as crack formation or moisture. These sensors send real-time data to a management platform, allowing engineers and technicians to assess the condition of the concrete on an ongoing basis. By combining this information with the concrete's self-repair capabilities, the repair process can be triggered at the right time, ensuring that cracks are closed before they become major problems (Henkensiefken & Schlangen, 2015).

One of the main advantages of this approach is improved concrete life cycle management. Monitoring systems allow for proactive, rather than reactive, maintenance, which means that interventions can be made before cracks worsen. This not only prolongs the service life of structures, but also reduces repair and maintenance costs in the long term. In addition, the integration of monitoring technologies can increase the confidence of owners and authorities in the safety of infrastructure, especially in critical applications such as bridges and high-rise buildings.

However, the implementation of self-repairing concrete with monitoring systems also presents challenges. One of them is the complexity of installation and maintenance of the monitoring systems, which require significant upfront investments. In addition, it is necessary to ensure that the sensors are compatible with the concrete and do not adversely affect its mechanical properties. The durability and reliability of monitoring devices under adverse conditions are also aspects that need to be evaluated (Tittelboom & De Belie, 2013).

In conclusion, self-healing concrete with monitoring systems represents a significant advance in materials engineering. This approach not only improves the repairability of concrete, but also optimises the management of its condition over time.

As technology advances and current challenges are overcome, it is likely that this synergy between self-repair and monitoring will become a standard in construction, contributing to safer and more sustainable structures.

1.4 Advantages of Self-Repairing Concrete

Improved Durability

One of the most significant advantages of repairable concrete is its ability to extend the service life of structures. By automatically sealing cracks that form due to stress or exposure to adverse environmental conditions, this type of concrete reduces the need for frequent and costly repairs. This not only improves the durability of structures, but also decreases the risk of structural failure (Al-Tabbaa & Azzam, 2015).

Maintenance Cost Reduction

The use of repairable concrete can result in a considerable reduction of maintenance costs in the long term. By minimising the need for manual interventions to repair damage, economic and human resources are saved. This is especially beneficial in critical infrastructure, where downtime and repair costs can be significant (Lepech & Li, 2008).

Environmental Sustainability

Self-repairing concrete contributes to environmental sustainability by reducing the amount of materials needed for repairs. By avoiding demolition and reconstruction of damaged sections, material waste is reduced and the carbon footprint associated with the production of new building materials is reduced. In addition, some types of repairable concrete use bacteria that are biodegradable and environmentally friendly (Jonkers & Schlangen, 2007).

Improving Structural Safety

The ability of repairable concrete to self-seal cracks contributes to greater structural safety. By preventing damage propagation, the risk of collapse or structural failure that could endanger people's lives is reduced. This is especially critical in buildings and bridges that carry heavy loads or are exposed to extreme weather conditions (Henkensiefken & Schlangen, 2015).

Versatility in Applications

Repairable concrete is versatile and can be applied in a variety of contexts, from residential buildings to road infrastructure and bridges. Its ability to adapt to different conditions and design requirements makes it an ideal choice for architects and engineers seeking innovative and effective solutions for modern construction (Al-Tabbaa & Azzam, 2015).

Increasing Efficiency in Construction

The implementation of repairable concrete can increase efficiency in construction processes. By reducing the need for subsequent repairs, projects can be completed more quickly and with fewer disruptions. This is particularly beneficial in large-scale projects where time is a critical factor (Lepech & Li, 2008).

1.5 Application of Self-Repairing Concrete

Self-repairing concrete is an innovation in materials engineering that has gained attention in recent decades due to its ability to improve the durability and sustainability of concrete structures. This type of concrete has applications in a variety of areas, from building construction to road infrastructure and bridges. Some of the most relevant applications of self-repairing concrete are detailed below.

One of the most common applications of self-repairing concrete is in building construction. This material can be used in residential and commercial structures, where damage prevention is crucial to the safety and longevity of the building. By automatically sealing cracks that may form due to structural stress or environmental changes, self-healing concrete reduces the need for frequent and costly repairs (Van Tittelboom et al., 2010).

Self-repairing concrete is also applied in road and pavement construction. Road surfaces are subject to constant wear and tear due to traffic and weather conditions. By using self-repairing concrete, the service life of roads can be extended, minimising maintenance and improving road safety. This is especially important in areas with extreme climates, where cracks can form quickly (Henkensiefken & Schlangen, 2015).

Bridges are critical structures that require constant maintenance to ensure the safety of users. The application of self-repairing concrete in bridge construction can help prevent deterioration and structural failure. This type of concrete can seal cracks that, if left untreated, could compromise the integrity of the bridge. In addition, by reducing the need for repairs, traffic disruptions are minimised and maintenance costs are optimised (Lepech & Li, 2008).

Dams and other hydraulic structures are exposed to extreme conditions and can suffer damage due to water pressure and erosion. Self-repairing concrete can be used in these applications to ensure that cracks are automatically sealed, helping to maintain the safety and functionality of these critical structures. This is especially relevant in the context of climate change, where environmental conditions can be more unpredictable (Al-Tabbaa & Azzam, 2015).

Self-repairing concrete also has applications in the restoration of historic buildings. By using this material, the structural integrity of old buildings can be preserved while minimising the visual impact of repairs. This is essential to maintain the historical and cultural value of these buildings, while ensuring their longevity (Jonkers & Schlangen, 2007).

1.6 Challenges and Future of Self-Repairing Concrete

Self-repairing concrete has emerged as an innovative solution in construction, promising to improve the durability and sustainability of structures. This material, which has the ability to automatically seal cracks as they form, offers significant potential to transform the way infrastructure is designed and maintained. However, its large-scale implementation faces several challenges that must be addressed to ensure its future success.

Current Challenges

One of the main barriers to the adoption of self-repairing concrete is the cost of production. The incorporation of admixtures, such as bacteria or microcapsules, can significantly increase the price of concrete compared to conventional concrete. This may be an impediment to its use in construction projects where budget is a primary concern (Zhao et al., 2022). Although costs are expected to decrease with advancing technology and large-scale production, the initial investment required may discourage builders and developers.

The lack of standardised rules and regulations for self-healing concrete is another major challenge. The construction industry is highly regulated, and the introduction of new materials requires validation of their performance and safety. Without clear standards, engineers and architects may be reluctant to adopt this technology (Zheng et al., 2021). The creation of specific regulations for self-repairing concrete could facilitate its acceptance and encourage its use in infrastructure projects, but this process can be slow and complicated.

The effective implementation of self-repairing concrete requires a change in the mindset of construction professionals. Many engineers and architects are still unfamiliar with the properties and benefits of this material. Therefore, it is essential to provide training and education on its use and application (Kaur & Singh, 2023). Without a clear understanding of how self-repairing concrete works, it is unlikely that professionals will use it in their designs. Training initiatives and educational programmes can be key to expanding its use in the industry.

Long-term Durability

Although self-healing concrete has been shown to be effective in laboratory tests, its long-term performance in real conditions still needs to be evaluated. Environmental conditions, such as exposure to moisture and dryness cycles, can affect the effectiveness of the self-healing mechanisms (Ravikar et al., 2023). In addition, it is crucial to assess how the ageing of the material influences its ability to self-heal over time. Continued research in this area is essential to ensure its long-term viability, especially in challenging environments.

Self-repairing concrete has the potential to reduce material waste and extend the service life of structures, but the production of the necessary admixtures can have a significant environmental impact. It is crucial to assess the full life cycle of self-repairing concrete to ensure that its benefits outweigh its environmental disadvantages (Tittelboom & De Belie, 2013). This includes considering the production and disposal of admixtures and how it compares to conventional concrete. Sustainability is an increasingly important factor in modern construction, and any new technology must be aligned with these goals.

Variability in the behaviour of self-repairing concrete can also make it difficult to implement. Different mix proportions, types of admixtures and curing conditions can influence the effectiveness of the material. Inconsistency in the quality of self-repairing concrete can lead to distrust in its use (İpek et al., 2023). Therefore, it is essential to establish testing and quality control procedures that ensure consistent performance under various conditions. Research on how to optimise these variables is essential to develop a reliable product.

Market acceptance is another critical factor affecting the implementation of self-repairing concrete. Resistance to change in the construction industry, combined with a lack of information on the benefits of self-repairing concrete, can hinder its adoption. It is essential to demonstrate its advantages in real projects and provide evidence of its efficacy and long-term economic benefits (Lesovik et al., 2021). Promoting successful case studies can help overcome these barriers and build confidence in the material.

Despite these challenges, the future of self-repairing concrete is promising. As technology advances, new solutions are being developed that could overcome current obstacles.

Investment in research and development is essential to improve the properties of self-healing concrete. New approaches, such as the use of nanomaterials and biological admixtures, are being explored to increase efficiency and reduce production costs. Collaboration between universities, research institutes and industry will be key to driving these advances (Lesovik et al., 2021). In addition, research into new curing methods and mix optimisation can improve the durability and self-healing ability of concrete. Collaborative networking will be essential to maximise the potential of these innovations.

As more studies demonstrate the effectiveness of self-healing concrete, it is likely that specific standards and certifications will be developed. This will facilitate its adoption in construction projects and increase practitioners' confidence in its use (Zheng et al., 2021). The creation of a clear regulatory framework will also incentivise companies to invest in this technology. Certifications can serve as a seal of approval, allowing builders and project owners to feel more confident in opting for this type of concrete.

Raising awareness of the benefits of self-repairing concrete is essential. Education and training programmes for engineers, architects and builders can help spread knowledge about this material and its applications, encouraging its use in future projects (Kaur & Singh, 2023). Initiatives such as workshops, seminars and conferences can be effective in informing professionals about advances in self-repairing concrete technology. In addition, outreach through media and digital platforms can help reach a wider audience, including students and new professionals.

Integrating self-healing concrete with emerging technologies, such as artificial intelligence and real-time monitoring, can improve its performance. For example, sensor systems could detect cracks and trigger repair mechanisms automatically, thus optimising the durability of structures (Ravikar et al., 2023). This combination of technologies would not only increase the efficiency of self-healing concrete, but also provide valuable data for analysis and continuous improvement. Digitalisation in construction promises to revolutionise the way infrastructure is managed and maintained.

As the construction industry moves towards more sustainable practices, self-repairing concrete can play a crucial role in the circular economy. Its ability to extend the lifespan of structures and reduce the need for new materials aligns with global sustainability goals (Zhao et al., 2022). Furthermore, the use of recycled and sustainable materials in the production of self-repairing concrete can contribute to a further reduction of the carbon footprint of construction. The search for sustainable solutions is becoming an imperative in modern construction, and self-repairing concrete can be an integral part of this transformation.

The development of new admixtures that improve the properties of self-healing concrete is a promising avenue for its future. Research is underway to find materials that not only improve self-healing capacity, but are also more economical and sustainable. For example, some studies are exploring the use of industrial by-products as admixtures, which could reduce environmental impact and production costs (İpek et al., 2023).

CAPÍTULO 2

Nanotecnología

2.1 Introduction

Nanotechnology has emerged as a transformative force in the construction industry, revolutionising the way materials are designed and manufactured. This technology is based on the manipulation of matter at the nano-scale, allowing the properties of materials to be modified to improve their performance and functionality. The applications of nanotechnology in construction are wide-ranging, from improving the strength and durability of materials to creating innovative solutions to environmental problems.

As the demand for more sustainable and efficient buildings continues to grow, nanotechnology presents itself as a viable answer to these challenges. Nanomaterials, with their unique properties, enable the creation of structures that are more resilient, durable and adaptive. For example, the use of nanocomposites in concrete not only improves its mechanical strength, but can also increase its durability against adverse environmental conditions (Vega-Baudrit & Juárez-Moreno, 2024). In addition, advanced coatings, which incorporate nanometric technology, offer features such as self-cleaning and corrosion resistance, prolonging the service life of buildings and reducing maintenance costs.

The implementation of nanotechnology in construction offers revolutionary potential, although it faces key challenges such as the lack of specific regulations and the need for further research into the safety of nanomaterials (Roco, 2005). This chapter will examine everything from the fundamental properties of nanotech materials to their specific applications in concrete, carbon nanotubes, and nanocomposites, as well as in the development of self-cleaning and durable materials. It will also explore the role of nanosensors for monitoring structures and analyse the environmental impact of these innovations.

2.2 Properties and Characteristics of Nanotechnological Materials

Nanotechnology has enabled the development of materials with unique properties that are not achievable with conventional materials. These properties are the result of the manipulation of matter at the atomic and molecular level, which allows the optimisation of physical, chemical and mechanical characteristics.

Nanotechnological materials are defined by their size, which varies between 1 and 100 nanometres. At this scale, materials exhibit properties that are drastically different from their macro-scale counterparts due to phenomena such as the quantum effect and the higher surface-to-volume ratio. The quantum effect causes electronic properties to behave differently, while the higher surface-to-volume ratio means that surface properties play a more important role, as seen in silver nanoparticles, which are highly effective as antimicrobials (Cámara Argentina de la Construcción, 2019).

Among the most outstanding mechanical properties of nanomaterials are their high strength and flexibility. For example, carbon nanotubes have tensile strengths exceeding 130 GPa, making them stronger than steel but much lighter (INSST, 2023). This combination of strength and flexibility is ideal for applications in structures that require both durability and the ability to absorb stress. In addition, some nanomaterials exhibit hardness comparable to diamond, opening up new possibilities for protective coatings in extreme environments (Repsol, 2023).

In terms of electrical and thermal conductivity, nanotechnological materials exhibit outstanding properties. Graphene is known for its exceptional electrical conductivity, with electron mobility that can reach up to 200,000 cm^2/V-s (UPM, 2017), making it a promising material for advanced electronic applications. Its high thermal conductivity, close to 5000 W/mK, enables the development of new, more efficient and compact electronic devices (INSST, 2023).

The optical properties of nanomaterials are equally fascinating and have significant practical applications. Nanoparticles can be designed to absorb or emit light at specific wavelengths; for example, gold nanoparticles can change their colour depending on their size and shape due to a phenomenon known as plasmon resonance (Cámara Argentina de la Construcción, 2019). In addition, titanium dioxide (TiO2)-based coatings are used in self-cleaning applications thanks to their photocatalytic ability to break down pollutants under UV light.

The practical applications of these materials are vast in the construction sector. The incorporation of nanoparticles such as silica or graphene into concrete significantly improves its mechanical characteristics and durability. This results in structures that are more resistant to adverse environmental conditions and reduces the risk of cracking (Cámara Argentina de la Construcción, 2019). Nanomaterials are also used to develop more efficient thermal insulation, significantly reducing energy losses in buildings (Repsol, 2023). In addition, the integration of nanomaterial-based sensors allows continuous monitoring of structural condition, detecting subtle changes that could indicate problems before they become serious failures (INSST, 2023).

With their ability to offer high mechanical strength, superior conductivity and advanced optical properties, these materials not only improve structural performance, but also promote more sustainable and innovative practices. As research in this field advances, we are likely to see even more transformative applications that positively impact the way we build.

2.3 Carbon nanotubes

Carbon nanotubes (CNTs) are emerging as an innovative material in civil construction, thanks to their exceptional mechanical properties and their ability to improve the durability and strength of building materials. These tubular structures, which can be single or multi-walled, have a diameter in the order of nanometres and are known for their high tensile strength, flexibility and electrical and thermal conductivity. Their incorporation into composites such as cement has been shown to significantly improve the mechanical properties of these materials, even in small quantities. For example, recent research has shown that the addition of functionalised carbon nanotubes can increase the strength of cement by up to 170% (European University, 2023).

Figure 2
Illustration of a carbon nanotube

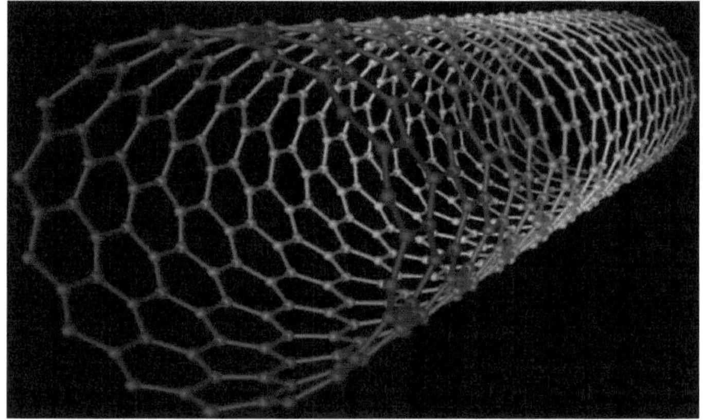

Source: Nanotec

Functionalisation of nanotubes is a crucial aspect of maximising their effectiveness in cement blends. Studies have indicated that functionalised nanotubes, which have specific chemical groups on their surface, interact better with cement, resulting in faster hydration and a 20% improvement in the strength of the material (European University, 2023). This interaction is due to the electrostatic attraction between the polar groups on the nanotube surface and the cement components. In addition, their ability to fill voids in the cement microstructure contributes to the reduction of porosity and improves the overall mechanical properties (Marcondes, 2012).

Carbon nanotubes also offer other interesting applications in civil construction. For example, they are being used to develop smart structures that can monitor their own condition. Researchers have proposed integrating nanotubes into beams and other structural elements to create communication channels that enable continuous monitoring of structural health (Agencia SINC, 2023). This technology could revolutionise infrastructure maintenance by enabling more accurate and less costly assessment of deterioration over time.

In addition, nanotubes are used to create self-healing materials and sensors that respond to environmental stimuli. This includes the possibility of developing composites that are not only resistant to adverse conditions, but can also adapt to changes in temperature or humidity (Dooko, 2023). These characteristics make nanotubes ideal for applications where high performance is required under extreme conditions.

However, the incorporation of carbon nanotubes in concrete mixtures presents challenges. The homogeneous distribution of these nanomaterials in the mix is crucial to maximise their benefits. Techniques such as ultrasound are being explored to achieve adequate dispersion of nanotubes in aqueous or oily solutions, which facilitates their effective integration into the materials (Marcondes, 2012). Despite these technical challenges, the potential of carbon nanotubes to transform civil construction is significant.

2.4 Nanoparticles in Concretes and Cements

Nanoparticles are transforming concretes and cements, bringing significant improvements in their mechanical and functional properties. These particles, which have dimensions on the nanometre scale (less than 100 nanometres), are integrated into cement mixes to modify their microstructure and improve their performance. The addition of nanoparticles such as nano silica (n.SiO2), nano titanium oxide (n.TiO2) and carbon nanotubes (CNT) has been shown to positively influence the cement hydration process at different scales: nano, micro and macro. These nanoparticles act as active nuclei that increase the hydration of cement, improving its strength properties, reducing its porosity and the concrete shrinkage that causes cracks, as well as its possible subsequent degradation (Expocihachub, 2023; Grinder, 2023).

Figure 3

Improving the microstructure of concrete using nanoparticles

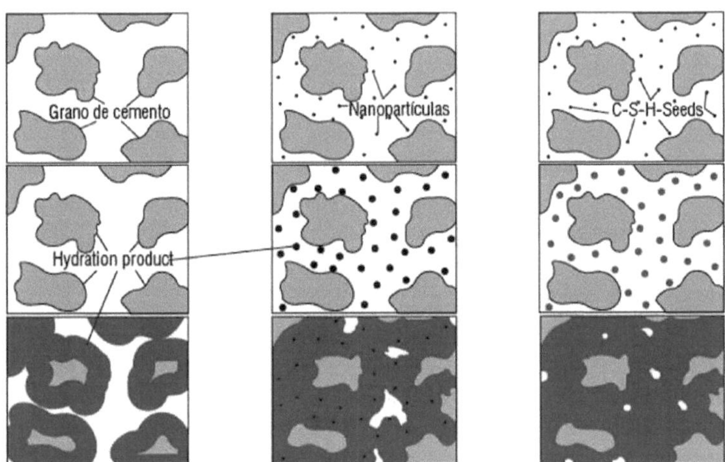

Source: Building better projects

2.4.1 Effects of Nanoparticles on Concrete Structure

The incorporation of nanoparticles in concrete has a noticeable impact on its microstructure. These particles fill the voids between the cement grains and between the aggregates, resulting in a higher density of calcium silicate hydrate (C-S-H) gel. This gel is crucial for the strength of the concrete; by increasing its density, the resistance to dissolution of the calcium carbonate present in the concrete matrix is increased. In addition, the presence of nanoparticles reduces the amounts of calcium hydroxide (Ca(OH)2) and the lower density C-S-H gel, which contributes to an overall improvement in the mechanical properties of the material (Grinder, 2023; Expocihachub, 2023).

2.4.2 Targeted Nanoparticles and their Benefits

- Nano Silica: Nano silica is especially effective in improving the strength and durability of concrete. Its high reactivity allows for better hydration of cement, resulting in a significant increase in compressive and flexural strength. It also helps to reduce the permeability of concrete, making it more resistant to chemical attack (Grinder, 2023).
- Nano Titanium Oxide: This material not only improves the mechanical properties of concrete, but also introduces photocatalytic characteristics that allow pollutants to be broken down under UV light. This means that concrete

treated with nano titanium oxide can contribute to air purification and self-cleaning of surfaces (Expocihachub, 2023).

- Carbon Nanotubes: As mentioned above, carbon nanotubes have proven to be particularly effective in improving the mechanical properties of cement. Research indicates that even small amounts of CNTs can increase the strength of cement by up to 170%. This is due to their ability to enhance hydration and increase the bond between concrete components (European University, 2023). In addition, nanotubes can be functionalised to optimise their interaction with cement, further increasing their effectiveness.

2.4.3 Innovations in Electrical Conductivity

A remarkable finding is that by adding black nanocarbon to the cement mix, concrete can be transformed into an electrical conductor. Research by MIT's Center for Concrete Sustainability has shown that with just 4% black nanocarbon, concrete can be made to carry electrical current. This opens up new opportunities for innovative applications such as integrated heating systems in interior floors (Expocihachub, 2023). By turning concrete itself into a heat conductor, the installation of heating systems is simplified and heat distribution is improved.

Researchers are also exploring the use of biodegradable nanoparticles such as chitin nanocrystals, derived from crustacean shells. These biopolymers not only improve the flexural and compressive strength of concrete by up to 40%, but also help reduce marine waste by using food by-products (Concrete a Day, 2023). This innovation not only improves the mechanical properties of concrete but also addresses environmental concerns by reducing the carbon footprint associated with its production.

2.4.4 Implementation Challenges

Despite the promising benefits, there are significant technical challenges in effectively incorporating nanoparticles into concrete mixes. Homogeneous distribution is crucial to maximise their benefits; methods such as ultrasound and chemical techniques to functionalise nanoparticles are being investigated to improve their dispersion (Marcondes, 2012). Furthermore, it is essential to understand how these nanoparticles interact with the resulting compounds during the cement hydration process.

2.5 Polymer Nanocomposites in Infrastructure

Polymer nanocomposites are emerging as an innovative solution in infrastructure construction, thanks to their improved properties and versatility in various applications. These materials are characterised by the homogeneous dispersion of nanometre-sized particles (smaller than 100 nm) within a polymer matrix, which gives them superior mechanical, thermal and electrical properties compared to conventional composites. The incorporation of nanoparticles, such as carbon nanotubes, graphene and nanoclays, has proven to be effective in improving the performance of polymers used in construction.

2.5.1 Mechanical and Thermal Properties

One of the most outstanding features of polymer nanocomposites is their improved mechanical properties. For example, studies have shown that adding carbon nanotubes (CNT) to polypropylene matrices significantly increases the impact strength and stiffness of the material. Research has shown that by increasing the NTC content, the mechanical properties of polypropylene improve significantly, allowing the development of lighter and more collision resistant materials (Redalyc, 2023). This type of reinforcement is crucial in applications where impact resistance is critical, such as structural components or coatings.

In addition, polymer nanocomposites have improved thermal properties. The uniform dispersion of nanoparticles within the polymer matrix not only increases stiffness, but also improves the thermal stability of the material. This is especially relevant in environments where temperatures can fluctuate drastically, as nanocomposites can maintain their structural and functional properties under extreme conditions (Zavala Murguía, 2023).

2.5.2 Enhanced Functionality

Nanocomposites not only improve mechanical and thermal properties; they also introduce new functionalities. For example, graphene has been integrated into polymer matrices to create materials with high electrical and thermal conductivity. This property is essential for applications where fast heat dissipation or efficient electrical conduction is required (SciELO, 2017). Graphene-based nanocomposites have proven particularly useful in electronic devices and solar energy systems, where their ability to conduct electricity can be harnessed.

In addition, self-cleaning coatings developed from polymer nanocomposites have gained attention due to their ability to reduce the maintenance required for infrastructure. These coatings use nanoparticles that allow the degradation of contaminants under UV light, resulting in surfaces that stay cleaner for longer (CIMAV, 2023).

2.5.3 Infrastructure Applications

Applications of polymer nanocomposites in infrastructure are diverse. They are used in the manufacture of composite materials for smart structures, where carbon nanotubes are integrated into beams and other structural elements to monitor their condition. This technology can detect structural changes or deterioration through embedded sensors that send information about the state of the material (Agencia SINC, 2023). This ability to monitor structural condition is crucial to ensure the safety and durability of large public works.

In addition, nanocomposites are being used in coatings for corrosion protection. The incorporation of metallic or ceramic nanoparticles can significantly improve the corrosion resistance of structures exposed to aggressive environments, thus prolonging their useful life (CIMAV, 2023).

2.5.4 Challenges and Future

Despite their numerous advantages, the implementation of polymer nanocomposites faces certain challenges. One of the main obstacles is ensuring a homogeneous dispersion of the nanoparticles within the polymer matrix. Agglomeration can significantly reduce the expected benefits. Therefore, advanced methods are being investigated to improve the dispersion and functionalisation of these nanoparticles (Zavala Murguía, 2023).

The future of polymer nanocomposites in infrastructure looks promising. With continued technological development and a greater understanding of how to manipulate these structures at the molecular level, we are likely to see an increase in their use not only in civil construction but also in other industries such as automotive and electronics.

2.6 Development of Self-Cleaning Materials

Nanotechnology has revolutionised the development of self-cleaning materials, offering innovative solutions that improve the cleaning and maintenance of various surfaces. These materials, which use nanoparticles to create unique properties, are designed to remove dirt and contaminants efficiently, reducing the need for manual cleaning and the use of harsh chemicals.

Self-cleaning materials are based on two key principles: superhydrophobicity and photocatalysis. Superhydrophobicity refers to the ability of a surface to repel water, allowing water droplets to roll off it, carrying dirt and other contaminants with them. This phenomenon is observed in nature, such as in lotus leaves, which have a surface structure that allows them to stay clean (StudySmarter, 2023).

Photocatalysis, on the other hand, involves the use of light (usually sunlight) to activate a component in the material, such as titanium dioxide (TiO2). When this material is exposed to light, it generates reactive species that can break down dirt particles and harmful microorganisms. This process not only keeps surfaces clean, but can also deactivate bacteria and viruses, which is especially useful in environments such as hospitals (CIMAV, 2023).

Recent research has led to the development of self-cleaning paints composed of coated titanium dioxide nanoparticles. These paints are applicable to a wide range of surfaces, including textiles, glass and steel. A study by researchers at University College London (UCL) has shown that these nanoparticles are not only effective under normal conditions, but also maintain their functionality even after being subjected to mechanical wear or exposure to oils (UCL Chemistry, 2023).

The self-cleaning paint created by UCL forms a finish that repels water and other liquids, creating a pearlescent effect on the surface. When water is applied to this treated surface, the droplets roll off and wash away the deposited dirt. In experiments on treated cotton, it emerged pristine white after being immersed in dyed water (Yao Lu et al., 2023). This type of technology has potential applications in a variety of sectors, from construction to automotive.

2.6.1 Applications in Various Sectors

Self-cleaning materials have a wide range of applications. In the construction sector, they are used for exterior and interior cladding that not only improve building aesthetics but also reduce maintenance costs by minimising the build-up of dirt and contaminants. For example, self-cleaning glass applied in skyscrapers helps to keep windows clean without the need for costly cleaning with scaffolding or cranes (StudySmarter, 2023).

In healthcare, self-cleaning paints can be used in hospitals to create surfaces that are not only easy to clean but also deactivate dangerous pathogens. This is crucial to prevent nosocomial infections and maintain clean and safe environments (CIMAV, 2023).

In addition, photocatalytic paints are being explored as solutions to combat air pollution. These paints can trap pollutant particles and break them down through chemical reactions triggered by ultraviolet light. Research by the Institute for Materials Chemistry in Vienna has shown promising results using recycled nanoparticles to create these paints (Imnovation Hub, 2023). The ability to purify the air and keep facades clean represents a significant advance towards more sustainable cities.

Despite the promising potential of nanotechnology-based self-cleaning materials, there are significant challenges to be addressed. One of the main obstacles is ensuring the durability and mechanical wear resistance of these materials after application. Although research has made progress in this area, it is crucial to continue to develop formulations that maintain their self-cleaning properties over time (Yao Lu et al., 2023).

The future of the development of self-cleaning materials looks bright. The market is expected to grow significantly in the coming years due to the increasing demand for sustainable and efficient cleaning solutions. The report "Markets for Self-Cleaning Coatings and Surfaces: 2015 to 2022" estimates market growth to reach approximately $3.3 billion by 2025 (UCL Chemistry, 2023).

2.7 Environmental Impact of Nanotechnological Materials

One of the main problems associated with nanotechnology is the potential toxicity of nanoparticles. Some nanoparticles have been shown to be more toxic per unit mass compared to their larger counterparts due to their larger surface area and chemical reactivity (The Royal Society, 2004). This raises concerns about how these particles may interact with living organisms and ecosystems. For example, it has been found that nanoparticles can be taken up by soil organisms, which could lead to adverse effects in the food chain (Redalyc, 2015).

In addition, the environmental fate of nanoparticles is a growing concern. It is estimated that a significant amount of generated nanoparticles end up in landfills or water bodies, where their behaviour and long-term effects are still poorly understood (Ve et al., 2015). The lack of data on the biodegradability and cumulative impact of these particles in the environment makes it difficult to assess risks and formulate appropriate regulations.

2.7.1 Sustainability in the Use of Nanotechnology

Despite the potential risks, nanotechnology also offers significant opportunities to improve sustainability in civil engineering. Self-cleaning materials based on nanotechnology, for example, can reduce the need for chemicals for cleaning and maintenance, thus minimising the associated environmental impact (CIMAV, 2023). In addition, photocatalytic coatings can help break down atmospheric pollutants, contributing to a cleaner urban environment (Rickerby et al., 2023).

The ability of nanomaterials to increase energy efficiency in buildings is also a key aspect to promote sustainable practices. For example, the use of nanocomposites in insulation systems can significantly reduce the energy consumption needed for heating and cooling (CIMAV, 2023). This not only decreases greenhouse gas emissions associated with energy use but also reduces long-term operating costs.

2.8 Challenges in the Application of Nanotechnology in Civil Engineering

Despite the promising potential of nanotechnology, its application in civil engineering faces several challenges:

- Lack of regulation: Regulating the use and production of nanotech products has been a challenge due to a lack of knowledge about the associated risks. Many nanotech products are considered new and are not covered by existing regulations (Nanobiology, 2023). This poses a risk to both workers handling these materials and the environment.
- Unknown Toxicity: The potential toxicity of nanoparticles remains an underexplored area. More ecotoxicological studies are needed to understand how these particles affect living organisms and ecosystems (Ve et al., 2015). Without this information, it is difficult to establish safe handling practices.
- Cost and Production: Large-scale production and the cost associated with nanotech materials can be prohibitive. Although costs are expected to decrease with technological advances, many projects currently face financial constraints that restrict their implementation (CIMAV, 2023).
- Public Awareness: There is a widespread lack of public understanding of the benefits and risks associated with nanotechnology. Education and awareness are crucial to foster informed public acceptance and facilitate the safe adoption of these technologies (Rickerby et al., 2023).
- Nanotechnology has the potential to revolutionise civil engineering through the development of more efficient and sustainable materials. However, it is critical to address the challenges associated with their application to ensure that they are used safely and responsibly. Appropriate regulation, together with further research into toxicity and environmental performance, will be essential to maximise the benefits while minimising the risks associated with these innovative materials.

Nanotechnology has the potential to revolutionise civil engineering through the development of more efficient and sustainable materials. However, it is critical to address the challenges associated with their application to ensure that they are used safely and responsibly. Appropriate regulation, together with further research into toxicity and environmental performance, will be essential to maximise the benefits while minimising the risks associated with these innovative materials.

Fuente: Cobod

CAPÍTULO 3

Impresión 3D

3.1 Introduction

3D printing has burst onto the construction industry as a revolutionary technology, transforming the way buildings are designed and built. This method of additive manufacturing allows three-dimensional structures to be created from digital models, depositing material layer by layer. The applications of 3D printing in construction are diverse, ranging from the production of prototypes and mock-ups to the construction of housing and infrastructure components.

One of the most attractive aspects of 3D printing is its ability to reduce costs and construction times. By minimising material waste and requiring less labour, this method can not only significantly decrease a project's budget, but also speed up its completion. This is crucial in emergency situations, where speed can be vital in providing shelter for communities affected by natural disasters.

In addition, 3D printing offers significant environmental benefits. The construction industry is responsible for a large proportion of carbon emissions, and adopting sustainable technologies is essential to mitigate this impact. 3D printing facilitates the use of recycled and sustainable materials, promoting greener practices.

However, its implementation faces challenges, such as the lack of specific regulations and resistance to change in a traditional sector. Despite these obstacles, the future of 3D printing in construction is promising. This chapter will explore in detail the basics of 3D printing, its current applications, the benefits and challenges it presents, as well as future trends that could redefine the construction industry.

3.2 Fundamentals of 3D Printing

3D printing, also known as additive manufacturing, has revolutionised the way objects are produced in a variety of industries, including construction. This process involves the creation of three-dimensional objects from digital models, using a variety of materials and techniques. 3D printing is based on the principle of adding material in successive layers, which contrasts with traditional manufacturing methods that often involve removing material, such as cutting or milling. This technique allows for greater flexibility in design and the creation of complex geometries that would be difficult to achieve with conventional techniques, as highlighted by Gibson, Rosen and Stucker (2015).

The 3D printing process generally follows several steps:

- Digital Model Creation: Computer-aided design (CAD) software is used to create a digital model of the object to be printed. This model is converted into a file in a format compatible with the 3D printer, such as STL or OBJ.
- File preparation: The digital file is prepared using slicing software, which divides the model into horizontal layers and generates the necessary instructions for the printer.
- Printing: The 3D printer uses the prepared file to deposit the material in layers, following the instructions of the cutting software. Depending on the technology used, the material can be extruded, sintered or cured.
- Post-Processing: Once printing is complete, the object may require post-processing, which may include sanding, painting or assembly of parts.

There are several methods of 3D printing, each with specific characteristics and applications. The most common is Fused Deposition Modeling (FDM), which uses thermoplastic filaments that are melted and extruded through a heated nozzle. This method is ideal for prototypes and functional parts due to its low cost and ease of use (Baker, 2019). Another important method is Stereolithography (SLA), which uses a liquid resin that is cured by an ultraviolet laser. SLA is known for its high precision and surface quality, which makes it suitable for applications that require fine details, such as jewellery and dental models (Chua & Leong, 2017).

Selective Laser Sintering (SLS) is a process in which a laser is used to sinter material powder, fusing the particles together to create a solid object. This method allows the creation of complex and functional parts and is common in the production of prototypes and final parts in industries such as automotive and aerospace (Gibson et al., 2015). Finally, Digital Light Processing (DLP) is similar to SLA, but uses a digital projector to cure resins, allowing layers to be printed more quickly, increasing production efficiency (Baker, 2019).

3D printing has a wide range of applications in various industries. In the field of rapid prototyping, it enables designers and manufacturers to create product prototypes quickly and cost-effectively, facilitating design iteration and improvement. In addition, 3D printing enables customised production, adapting to the specific needs of users, from medical prosthetics to jewellery (Chua & Leong, 2017). In construction, this technology is used to create architectural components and, in some cases, entire buildings, which can reduce construction time and costs.

Figure 4

3D printed bridge

Source: Arcus global

The adoption of 3D printing offers numerous benefits, including:

- Waste reduction: By using only the amount of material needed to create an object, this method generates less waste compared to traditional methods (Baker, 2019).
- Design Flexibility: The ability to create complex geometries without the limitations of conventional manufacturing methods allows designers to explore new ideas and concepts (Chua & Leong, 2017).
- Time and Cost Savings: 3D printing can significantly reduce production times and associated costs, especially in the creation of prototypes and custom parts (Gibson et al., 2015).

Despite its advantages, 3D printing faces challenges that need to be addressed. Material limitations represent an obstacle, as although new composites are being developed, the variety of options for 3D printing is still limited compared to traditional manufacturing methods (Baker, 2019). Also, the lack of specific regulations for 3D printing in certain industries may hinder its adoption and use in critical applications, such as medicine and construction (Chua & Leong, 2017). Finally, ensuring the quality and consistency of 3D printed products can be challenging, especially in mass production (Gibson et al., 2015).

3.3 Current Applications in Construction

3D printing has begun to transform the construction industry, offering innovative solutions that address traditional challenges such as cost, construction time and sustainability. As the technology advances, its applications are diversifying, enabling the creation of complex, customised structures that were previously difficult to achieve. Below are some of the most prominent applications of 3D printing in construction.

3.3.1 Housing Construction

One of the most notable applications of 3D printing in construction is the creation of homes. Companies such as Texas-based ICON have developed 3D printers capable of building houses in record time. In 2018, ICON unveiled its "Vulcan" model, which can print a house in approximately 24 hours using a special concrete material. This approach not only reduces construction time, but also significantly lowers costs, with estimates indicating that a house can cost as little as $10,000 (Khoshnevis, 2018).

Figure 5

House built by the company ICON from 3D printing

Source: Regan Morton Photography

In addition, the Russian company Apis Cor has carried out similar projects, building a house in just 24 hours at a construction site in Moscow. Using a giant 3D printer, Apis Cor applies a mixture of recycled materials and quick-drying cement, allowing for fast and efficient construction (Apis Cor, 2017). These initiatives not only address the need for affordable housing, but also offer solutions in emergency situations, such as natural disasters.

3.3.2 Infrastructure and Public Works

3D printing is also being used for the construction of infrastructure, such as bridges and canals. In 2016, the first 3D printed pedestrian bridge, designed by ACCIONA, was inaugurated in Madrid. This 12-metre-long bridge was built using a 3D printer that placed material only where it was needed, allowing for greater design freedom and a reduction in the use of materials (ACCIONA, 2016). Such projects demonstrate how 3D printing can be used to create durable and aesthetically pleasing structures while optimising resources.

Another example is the 3D printed canal in Amsterdam, which was created using a special 3D printer known as the Kamermaker. This device, which resembles a giant robotic arm, enabled the construction of a canal efficiently and accurately, showing the potential of 3D printing in the creation of urban infrastructure.

Figure 5

3D printed canal in Amsterdam's red light district

Source: Thijs Wolzak

3.3.3 Customised Architectural Components

3D printing enables the creation of customised architectural components, offering architects and designers unprecedented flexibility. For example, the architectural firm SOM, in collaboration with Oak Ridge National Laboratory, developed a 3D printed green house that incorporates solar panels and renewable energy systems. This project not only demonstrates the ability of 3D printing to create innovative designs, but also highlights its potential to contribute to sustainability in construction (SOM, 2019).

Customisation of architectural components also extends to decorative and functional elements, such as railings, columns and facades. The ability to custom print these elements allows designers to experiment with shapes and textures that would be difficult to achieve with traditional construction methods.

Figure 6

3D printed street furniture

Source: El español

3.3.4 Prototyping and Modelling

Rapid prototyping is another area where 3D printing has had a significant impact on construction. Architects and designers can create physical models of their projects in a reduced timeframe, making it easier to visualise and evaluate designs prior to actual construction. This not only saves time, but also reduces costs by allowing adjustments in the early stages of design (Khoshnevis, 2018).

In addition, 3D printing enables the creation of scale models that can be used for presentations to clients or for obtaining building permits. These detailed mock-ups help communicate the project vision more effectively, which can be crucial for project approvals in regulated environments.

3.3.5 Sustainability and Waste Reduction

Sustainability is a key aspect of 3D printing in construction. By using only the amount of material needed to create an object, this method generates less waste compared to traditional manufacturing methods. In addition, many companies are exploring the use of recycled and sustainable materials in their 3D printers, contributing to greener building practices (Apis Cor, 2017).

3D printing's ability to optimise material usage and reduce construction time also has a positive impact on the carbon footprint of construction projects. As the industry looks for ways to become more sustainable, 3D printing presents itself as a viable solution that can help achieve these goals.

3.4 Benefits of 3D Printing in Construction

3D printing has emerged as a revolutionary technology in a number of industries, and construction is no exception. This method of additive manufacturing offers multiple benefits that can transform the way structures are designed and built. Here are some of the most significant benefits of 3D printing in construction.

One of the most significant benefits of 3D printing in construction is cost reduction. This approach allows only the required amount of material to be used, minimising waste and thereby reducing the costs associated with purchasing and storing materials. Research indicates that 3D printing can reduce construction costs by 30-60% compared to traditional methods (Khoshnevis, 2018). Furthermore, by automating the process, labour costs can be significantly reduced, representing additional savings for contractors. The combination of these factors makes 3D printing an attractive option for construction projects, especially in a context where economy and efficiency are essential. This approach not only benefits builders, but also has the potential to make housing more affordable for low-income communities, thus contributing to solving the housing crisis in many areas.

3D printing allows construction projects to be completed in a significantly shorter time than conventional methods. While traditional construction can take months or even years, 3D printing can complete structures in a matter of days or even hours, depending on the complexity of the design. For example, some 3D printed houses have been built in less than 24 hours (Apis Cor, 2017). This speed not only benefits builders, but also allows homeowners to occupy their new homes more quickly, which is especially valuable in emergency situations or natural disasters. The ability to speed up the construction process not only improves efficiency, but can also result in a faster revenue stream for contractors, which is crucial in a competitive market. In a world where speed and efficiency are increasingly valued, 3D printing presents itself as a viable and attractive solution.

Sustainability is a fundamental aspect of modern construction, and 3D printing makes a significant contribution to this goal. This method generates less waste compared to traditional construction, where large amounts of waste are often produced due to cutting and leftover materials. 3D printing uses only the material needed, which can result in a reduction of up to 60% in waste generated at the construction site (Cemex Ventures, 2021). In addition, many companies are exploring the use of recycled and sustainable materials in their 3D printers, further contributing to the sustainability of the process. The ability to optimise the use of materials not only helps protect the environment, but can also lower the operating costs of projects. As the construction industry looks for ways to be greener, 3D printing is positioned as a viable solution that can contribute to more responsible building practices.

Workplace safety is a constant concern in the construction industry, and 3D printing can help to improve it. This method reduces the need for workers to perform dangerous tasks on the construction site. By automating the process, human interactions with heavy machinery are minimised and injury risks are reduced (Cemex Ventures, 2021). The use of 3D printers allows many tasks to be performed in controlled environments, which also reduces exposure to adverse weather conditions. Improved safety not only protects workers, but can also result in lower costs associated with workplace accidents. In an industry where safety is paramount, 3D printing presents itself as an alternative that not only optimises efficiency, but also prioritises the well-being of employees.

3.5 Challenges and Limitations

3.5.1 Material Limitations

3D printing has opened up new possibilities in construction, but one of the main challenges it faces is the limited range of materials available. Although specific concretes and plastics have been created for 3D printing, the range of options is still restricted compared to traditional construction methods. Currently, materials used in 3D printing must meet specific technical requirements, such as the ability to flow through the printing nozzle and cure properly to form solid structures.

Concrete used in 3D printing often contains special admixtures to improve its workability and setting time. However, these admixtures can compromise some mechanical properties, such as compressive strength or long-term durability. This raises concerns about the quality and safety of structures built with these materials. Research and development of new materials specifically for 3D printing requires significant effort in terms of time, funding and human resources, which limits the capacity for innovation in this field (Khoshnevis, 2018).

In addition, available materials must be compatible with 3D printers and their technologies. For example, most construction 3D printers use extrusion processes, which require the material to have adequate rheological properties to flow and be modelled properly. This means that not all types of concrete or plastics can be used without modification. The need to develop new composites that maintain strength and durability while being suitable for 3D printing is a constant challenge.

Another aspect to consider is the sustainability of materials. The construction industry is facing increasing pressures to reduce its carbon footprint and use more environmentally friendly materials. Most of the concretes used in 3D printing are derived from non-renewable resources, raising questions about their long-term sustainability. The search for alternative materials, such as recycled or biomass-based concretes, is critical for 3D printing in construction to align with global sustainability initiatives.

Finally, research on the long-term behaviour of 3D printed materials is still in its infancy. More studies are needed to understand how these materials respond to ageing, exposure to adverse environmental conditions and mechanical factors such as loading. The lack of long-term data on the durability of these materials may limit confidence in their use for permanent buildings.

In summary, while 3D printing has the potential to transform construction, material limitations are a significant challenge that needs to be addressed. The need to develop new composites that are suitable for printing, sustainable and meet quality and durability standards is crucial for the future of this technology in the construction sector.

3.5.2 Standards and Regulations

The lack of clear standards and regulations is a critical challenge facing 3D printing in construction. The construction industry is highly regulated to ensure the safety, quality and integrity of buildings. However, the introduction of emerging technologies such as 3D printing raises questions about how these regulations apply to new construction methods.

Existing regulations are often not designed to address the specificities of 3D printing. For example, regulations governing structural strength, fire safety and energy efficiency may not address additive manufacturing methods. This can lead to legal uncertainties and resistance from regulators, who may be reluctant to approve projects that use unproven or poorly understood technologies (PlanRadar, 2021).

In addition, the lack of specific standards for materials used in 3D printing can further complicate the situation. Without clear standards, contractors and architects may face difficulties in ensuring that the materials used meet the necessary requirements for safe construction. This can result in a lack of confidence in the technology on the part of clients and authorities, which in turn can slow down the adoption of 3D printing in construction projects.

Another aspect to consider is the variability in regulations in different regions and countries. In some places, 3D printing in construction may be more advanced and have a more developed regulatory framework, while in others, it may be completely absent. This disparity can create an uneven competitive environment, where companies operating in regions with more flexible regulations can advance more quickly than those in areas where regulations are more restrictive.

To address these challenges, it is essential that industry collaborates with regulatory bodies to develop regulations that accommodate 3D printing technologies. This includes the creation of standards that assess the safety, quality and performance of printed structures. Collaboration between engineers, architects, material manufacturers and regulators is essential to establish a regulatory framework that encourages innovation while ensuring safety and quality in construction.

In conclusion, the lack of clear standards and regulations is a significant barrier to the adoption of 3D printing in construction. Addressing this challenge is crucial to allow this technology to evolve and be implemented safely and effectively in the sector, contributing to a more innovative and sustainable future in construction.

3.5.3 Initial Costs and Accessibility

While 3D printing promises to reduce long-term costs in construction, the initial investment costs in technology and equipment can be prohibitive for many companies, especially small and medium-sized ones. Large-scale 3D printers, along with specialised materials, require a significant investment that may not be feasible for all contractors. This financial challenge limits the adoption of this innovative technology in construction.

Construction-specific 3D printers are expensive, and the cost includes not only the purchase of the equipment, but also its installation, maintenance and operation. Companies must consider training staff to operate these machines, as well as learning the specialised software needed to design and prepare models for printing. This training represents an additional cost that many companics may not bc willing or able to afford (Cemex Ventures, 2021).

In addition, the costs of materials used in 3D printing can also be higher than conventional materials. While it is expected that, over time, the production of 3D printing-specific materials will be optimised and costs reduced, many of these materials are currently more expensive due to their specialised nature and the manufacturing processes required.

The initial investment can deter small and medium-sized enterprises from exploring 3D printing, which in turn limits competition and innovation in the sector. Without access to this technology, these companies may miss opportunities to improve their construction processes and offer more efficient and sustainable solutions.

To overcome this obstacle, it is essential that business models are developed that allow companies to share resources and reduce costs. For example, companies could consider partnerships to purchase 3D printers and materials, or explore financing options that facilitate initial investment. In addition, research and development initiatives funded by government or industry partnerships can help reduce the financial burden of adopting this technology.

It is also essential to raise awareness of the long-term benefits of 3D printing in construction. While initial costs may be high, the reduction in labour and material costs in the future, along with the ability to minimise waste, can result in significant savings over time. By educating companies about these benefits, greater acceptance and exploration of 3D printing in the construction sector can be encouraged.

In summary, upfront costs and accessibility are significant challenges to the adoption of 3D printing in construction. Overcoming these obstacles will require a collaborative approach involving strategic partnerships, innovative business models and increased education on the long-term benefits of this technology.

3.5.4 Scalability

Scalability is one of the most relevant challenges in the application of 3D printing in construction. While significant progress has been made in creating small and medium-sized structures, printing large-scale buildings remains a considerable challenge. As projects become larger and more complex, 3D printers are required to handle large volumes of material and operate efficiently, posing technical and logistical challenges.

Construction 3D printers must be able to print large quantities of material in a reasonable amount of time, which can be complicated. Print speed is a critical factor; if printers cannot produce components at an adequate rate, this can lead to project delays and increased costs. In addition, the logistics of large-scale 3D printing involve the need for significant space on the construction site, as well as the coordination of material delivery and the management of drying and curing time (Khoshnevis, 2018).

Another aspect of scalability is the ability of printers to adapt to different designs and specifications. Customisation of designs is one of the advantages of 3D printing, but this can also complicate the large-scale production process. Each design may require adjustments to the configuration of the printer and the type of material used, which can affect the overall efficiency of the process.

Integrating 3D printing into the existing construction supply chain also presents challenges. Transitioning from traditional construction methods to 3D printing requires changes in project planning and execution, which can be complicated for companies that have been using conventional techniques for years. Lack of experience in managing 3D printing projects can lead to errors and delays, which reinforces the dominance of traditional methods in large-scale projects.

To address these challenges, it is essential that 3D printing technologies are developed that are more efficient and adaptable to large-scale projects. This may include the creation of larger 3D printers that can handle larger volumes of material and have the capacity to operate faster and more effectively. It is also important to foster collaboration between industry players to share knowledge and best practices in large-scale 3D printing projects.

In conclusion, scalability in 3D printing is a critical challenge that must be overcome for this technology to compete effectively with traditional construction methods on large-scale projects. As innovative and efficient solutions are developed, 3D printing is likely to become a viable and effective option for a wider range of construction applications.

3.5.5 Market Perception

Market perception of 3D printing in construction can be a significant barrier to adoption. Many industry professionals and potential clients may be sceptical about the quality and safety of 3D printed structures. This mistrust may be based on a lack of successful examples and limited experience in the use of this technology, leading to doubts about its feasibility and effectiveness (PlanRadar, 2021).

Negative perceptions can be fuelled by a lack of clear and accessible information about the benefits and capabilities of 3D printing. Many industry players may be unfamiliar with the technical aspects of 3D printing, which can lead to misunderstandings about its applicability and performance. To overcome this challenge, education and awareness campaigns that inform construction professionals about the benefits of 3D printing, as well as success stories in its implementation, are essential.

The lack of tangible examples of successful projects also contributes to the negative perception. As more construction projects are completed using 3D printing, it is crucial to document and share these success stories. Presenting case studies that demonstrate the quality, safety and efficiency of 3D printed structures can help build confidence in the market and encourage the adoption of this technology.

In addition, collaboration with regulatory and standards bodies can help increase confidence in 3D printing. If regulatory authorities recognise and approve 3D printed construction projects, this can help validate the technology and reduce the perception of risk associated with its use.

Another important aspect is the relationship between architects, engineers and contractors. Communication and collaboration between these professionals is essential to ensure that the capabilities and limitations of 3D printing are understood. By working together on projects that incorporate this technology, they can share knowledge and experience, which can help build a foundation of trust and encourage greater acceptance of 3D printing in the marketplace.

In short, market perception of 3D printing in construction can be a significant barrier to adoption. To overcome this challenge, it is critical to educate industry professionals, share success stories and foster collaboration between industry players. Over time, as 3D printing demonstrates its effectiveness and safety in real projects, market perception is likely to evolve, opening the door to greater acceptance and use of this technology in construction.

3.5.6 Technical Challenges

The technical challenges associated with 3D printing in construction are varied and complex. Accuracy in printing, quality control and the management of material curing and drying times are critical aspects that require careful attention. Each of these factors can significantly influence the final outcome of a construction project and, if not properly managed, can result in structural defects that compromise the integrity of the building (Cemex Ventures, 2021).

Accuracy in printing is critical to ensure that structures comply with the specified designs. Deviations in dimensions and geometry can lead to alignment and fit problems during assembly. This becomes especially problematic in large projects, where dimensional tolerance is crucial for structural stability. It is therefore essential to ensure that 3D printers are calibrated correctly and that quality control techniques are used throughout the printing process.

Quality control is also a vital aspect of 3D printing. Since the printing process is done in real time, it is difficult to monitor each step to ensure that there are no defects. Implementing real-time monitoring systems that can detect anomalies during the printing process can help mitigate this challenge. In addition, quality testing of the materials used, as well as the printed components, is essential to ensure that safety and performance standards are met.

Managing the drying and curing times of materials is another significant technical challenge. Different materials require different curing conditions to achieve their optimum properties. Variability in environmental conditions, such as temperature and humidity, can affect the performance of materials, further complicating the process. Lack of control over these conditions can lead to the formation of cracks and other structural defects, which in turn can compromise the durability of the final structure.

In addition, the complexity of architectural designs that can be achieved through 3D printing can pose technical challenges. While the technology enables the creation of innovative forms and structures, these complexities may require greater technical knowledge and design expertise to ensure that the structures are viable and safe.

In summary, the technical challenges in 3D printing for construction are manifold and require a meticulous and well-planned approach to ensure successful projects. The implementation of quality control systems, careful management of curing times and a focus on accuracy are essential to overcome these obstacles and ensure that 3D printing is used effectively in construction.

3.5.7 Integration with Traditional Construction Methods

The integration of 3D printing with traditional construction methods represents a significant challenge in the adoption of this technology. The transition to adopting new technologies requires changes to existing processes, which can be complicated for companies that have been using conventional methods for years. Resistance to change is a common phenomenon in the construction industry, where familiarity with traditional methods can make it difficult to accept innovations such as 3D printing (PlanRadar, 2021).

One of the main barriers to integration is the lack of interoperability between design tools used in traditional construction and 3D printers. Many companies use specific software for building planning and design that may not be compatible with 3D printing platforms. This lack of compatibility can lead to errors in data transfer and complicate the design and printing process, making it difficult to effectively implement 3D printing in projects.

In addition, construction companies may lack the expertise to manage projects that incorporate 3D printing. Lack of staff trained in operating 3D printers and understanding additive manufacturing processes can limit the ability of companies to adopt this technology effectively. Staff training is essential to ensure that the capabilities and limitations of 3D printing are understood, as well as to maximise its potential in construction projects.

Another challenge is the need to modify existing supply chains. 3D printing can reduce the need for prefabricated materials and components, which can affect suppliers that rely on traditional manufacturing. Adapting supply chains to a 3D printing approach will require the collaboration and engagement of multiple stakeholders in the construction industry.

To facilitate the integration of 3D printing with traditional construction methods, it is essential to develop solutions that promote interoperability between different platforms and tools. This could include the creation of data standards and protocols that facilitate communication between different parts of the construction process. In addition, fostering collaboration between architects, engineers, contractors and suppliers can help build a deeper understanding of how 3D printing can be used in combination with traditional methods.

In conclusion, the integration of 3D printing with traditional construction methods represents a complex challenge that needs to be addressed to enable wider adoption of this technology.

3.6 The Future of 3D Printing in Construction

3D printing has emerged as a revolutionary technology in construction, promising to transform the way buildings are designed and built. This process, also known as additive manufacturing, allows the creation of complex structures by layering material, offering a number of advantages over traditional construction methods. As the technology advances, new opportunities and challenges are emerging that will define the future of 3D printing in construction.

One of the most prominent aspects of 3D printing in construction is its ability to reduce costs and production times. According to a recent study, 3D printing can significantly decrease construction time, allowing structures to be erected in a matter of days rather than months (Khoshnevis, 2018). This not only speeds up the construction process, but also reduces labour and material costs, which can make housing more affordable for many people.

In addition, 3D printing allows for greater customisation in architectural design. Architects and designers can create forms and structures that would be difficult or impossible to achieve with conventional construction methods. This translates into greater creative freedom and the ability to design buildings that integrate better with their surroundings (Garcia et al., 2020). The ability to produce customised components can also improve the energy efficiency of buildings, as elements can be designed to optimise natural light and ventilation.

However, despite its many advantages, 3D printing in construction faces several challenges. One of the main obstacles is regulation. The construction industry is highly regulated, and the introduction of new technologies such as 3D printing requires the adaptation of existing regulations. This can slow down the adoption of the technology, as legislators must ensure that printed structures meet safety and quality standards (Zhang, 2021). In addition, the lack of clear standards for materials used in 3D printing can lead to uncertainty in the market.

Another significant challenge is the need for specialised training. 3D printing in construction requires not only technical knowledge about the operation of the printers, but also a thorough understanding of the materials and their behaviour. The training of skilled professionals in this technology is essential for its successful implementation in construction projects. However, education in this field is still developing, which may limit the availability of skilled labour (Cemex Ventures, 2021).

As technology advances, new applications of 3D printing in construction are also being explored. For example, techniques are being developed to print structures using sustainable materials, such as bioplastics and recycled concrete. This not only reduces the environmental impact of construction, but also promotes the circular economy by reusing materials that would otherwise be discarded. Integrating 3D printing with other emerging technologies, such as artificial intelligence and robotics, could also improve efficiency and accuracy in construction.

In summary, the future of 3D printing in construction is promising, with the potential to significantly transform the industry. As current challenges are overcome and new applications are developed, we are likely to see an increase in the adoption of this technology. The combination of cost reduction, design customisation and sustainability could make 3D printing an attractive option for the future of construction.

CAPÍTULO 4

BIM

4.1 Introduction to BIM

Building Information Modelling (BIM) is a revolutionary approach in architecture, engineering and construction that transforms the way buildings and infrastructure are designed, built and managed. In essence, BIM refers not just to a software or tool, but to a collaborative process that integrates and manages information throughout the life cycle of a project. From initial planning to demolition, BIM provides a detailed digital model that includes not only the physical geometry of the building, but also data on materials, costs, schedules and energy performance.

Figure 7

Source: Steel study.

The evolution of BIM has been remarkable. Although the concepts of three-dimensional modelling in construction can be traced back to the 1970s, it was in the 1990s that the term 'BIM' began to gain popularity. Initially, it focused on the creation of static 3D models; however, with the advancement of technology, BIM has evolved into a dynamic and collaborative environment, where multiple disciplines can work simultaneously on a single model. This advance has been driven by the development of more sophisticated software and the growing need to improve efficiency and reduce costs in an industry known for its complexity.

A comparison between traditional design and construction methods and the BIM approach highlights several key differences. In traditional methods, teams often work in silos, meaning that architects, engineers and contractors often develop their designs separately. This can lead to communication problems, design conflicts and, ultimately, delays and cost overruns. Two-dimensional drawings are often misinterpreted, which can result in costly errors during construction.

In contrast, the BIM approach encourages collaboration from the outset. All stakeholders can access a centralised model, allowing them to visualise the project in three dimensions and make modifications in real time. This not only improves communication, but also facilitates the detection of interferences before construction starts. In addition, the use of integrated data allows for more effective planning and more accurate building lifecycle management.

In short, BIM represents a radical change in the way construction is approached, offering a more collaborative, efficient and sustainable approach compared to traditional methods. With its ability to integrate information and facilitate communication, BIM is transforming the industry, preparing it to meet the challenges of the future.

4.2 Components of BIM

Building Information Modelling (BIM) is a methodology that has transformed the way buildings and other infrastructure are designed, constructed and managed. This methodology is based on the creation of digital models that integrate both the geometry and the information associated with the elements of the project. The following will explore the components of BIM, the importance of 3D models, the data linked to the elements of the model and the software tools most commonly used in this methodology.

The components of BIM include 3D models, data and information linked to the model elements, and software tools that facilitate the creation and management of these models. Each of these components plays a crucial role in the life cycle of a construction project.

4.2.1 3D Models and their Importance

3D models are the core of BIM. Unlike traditional 2D models, which only represent the geometry of a building, 3D models in BIM integrate additional information that is essential for project planning and execution. This information can include details on materials, costs, energy performance and more.

- Visualisation and Understanding: 3D models allow architects and other professionals to visualise the project in a three-dimensional environment. This not only improves understanding of the design, but also makes it easier to identify potential problems before construction begins. The ability to view the project from different angles and at different stages of development is critical for informed decision making (Eastman et al., 2011).
- Conflict Detection: One of the most significant advantages of 3D models is their ability to detect interferences between different systems, such as structures,

electrical and plumbing systems. The use of software that allows the overlay of models from different disciplines helps to identify and resolve design problems before they become costly errors during construction (Korman, 2010).

- Simulation and Analysis: 3D models also allow for simulations and performance analysis, such as lighting studies and energy analysis. These simulations are crucial to optimise the design and ensure that the building meets sustainability and energy efficiency standards (Azhar, 2011).
- Integrated documentation: 3D models in BIM not only serve as visual representations, but are also linked to the project documentation. This means that any changes made to the model are automatically reflected in associated documents such as drawings and specifications. This integration reduces the possibility of errors and ensures that all team members are working with the most up-to-date information (Sacks et al., 2010).

4.2.2 Data and Information Linked to Model Elements

BIM is not limited to visual representation; it also relies on the integration of data and information linked to each element of the model. This information is crucial for efficient project management and informed decision making.

Element Information: Each component in the 3D model is associated with specific data describing its characteristics, such as dimensions, materials, costs and performance. For example, a window in the model will not only be represented visually, but will also include information about its type, size, energy efficiency and cost. This wealth of data allows project teams to perform detailed analyses and make decisions based on accurate information (Bynum et al., 2013).

Life Cycle Management: The information linked to the elements of the model is fundamental to the management of the building's life cycle. From planning and construction to operation and maintenance, BIM provides a framework for managing all aspects of the project. Maintenance data on mechanical and electrical systems can be used to schedule interventions and optimise performance over time (Kiviniemi et al., 2012).

Collaboration and Communication: The ability to share detailed and up-to-date information among all team members is one of the most valuable features of BIM. Data linked to model elements facilitates collaboration between architects, engineers and contractors, ensuring that everyone is working with the same information and reducing the risk of misunderstandings and errors (Chong et al., 2017).

Cost Analysis and Budgeting: Integrating data into the model also allows for more accurate cost analyses. By linking material and quantity information directly to the model, teams can generate more reliable cost estimates and track budget throughout the project. This is especially useful in the planning phase, where accurate estimates can make the difference between project success and failure (Tzortzopoulos et al., 2011).

4.2.3 Most Used BIM Software and Tools

The use of specialised software is essential to implement BIM effectively. There are several tools on the market that enable construction professionals to create and manage BIM models. Some of the most widely used include:

1. Revit: Developed by Autodesk, Revit is one of the most popular tools for BIM modelling. It allows users to design with parametric drawing and modelling elements, facilitating the creation of detailed 3D models that integrate information about building elements. Revit also enables real-time collaboration between different disciplines, which improves project coordination (Eastman et al., 2011).

2. Navisworks: This tool is particularly useful for model review and coordination. Navisworks allows users to combine models from different disciplines in a single environment, facilitating clash detection and construction planning. Its ability to perform construction simulations and time analysis makes it a valuable tool for project management (Korman, 2010).

3. ArchiCAD: Developed by Graphisoft, ArchiCAD is another popular tool in the field of BIM. It offers an intuitive approach to architectural design and allows users to create 3D models efficiently. ArchiCAD also includes tools for collaboration and data management, making it a solid choice for design teams (Bynum et al., 2013).

4. Tekla Structures: This tool focuses on structural modelling and is widely used in civil engineering and construction. Tekla allows users to create detailed models of steel and concrete structures, integrating material and cost information. Its ability to manage the construction life cycle makes it especially valuable for complex projects (Azhar, 2011).

5. Dynamo: Although not a modelling software per se, Dynamo is an add-on for Revit that enables visual programming. Users can create custom algorithms to automate tasks and generate complex geometries, extending Revit's capabilities and improving design efficiency (Sacks et al., 2010).

4.3 Advantages of BIM

The Building Information Modelling (BIM) methodology has transformed the way construction projects are managed, offering significant advantages that impact on collaboration between disciplines, reduction of errors and conflicts, as well as time and cost savings throughout the project lifecycle.

One of the main advantages of BIM is the improved collaboration between disciplines. This methodology allows architects, engineers and contractors to work in a common digital environment, where information is shared and updated in real time. This collaborative approach not only facilitates communication, but also promotes more effective coordination between the different teams involved in the project. According to Eastman et al. (2011), working simultaneously on a digital model allows professionals to identify and solve problems early, which significantly reduces the risk of conflicts during the construction phase. The integration of different disciplines in a single model also encourages the creation of specific roles, such as the BIM coordinator, who is responsible for managing information and ensuring that all elements of the project are aligned.

Reducing errors and conflicts is another critical advantage of BIM. The ability to detect interferences between different building systems before construction begins is critical to avoid costly rework. Tools such as Navisworks allow for detailed analyses that identify potential problems, such as overlapping structural and facility elements, allowing teams to address these conflicts at the design phase rather than on the construction site (Korman, 2010). This early detection not only minimises errors, but also improves the quality of the final project, as adjustments and optimisations can be made before problems materialise.

In addition, the use of BIM contributes to time and cost savings throughout the project lifecycle. The ability to generate accurate estimates and track costs in real time allows project teams to better manage their resources. By linking cost information directly to the model, variances can be quickly identified and the budget adjusted as needed (Azhar, 2011). This is especially beneficial in the planning phase, where an accurate estimate can make the difference between project success and failure. In addition, process automation and reduced documentation errors also contribute to more efficient execution, resulting in reduced construction time and thus significant savings in operational costs.

Figure 8

BIM methodology in the life cycle of a building.

CICLO DE VIDA DE LA EDIFICACIÓN.

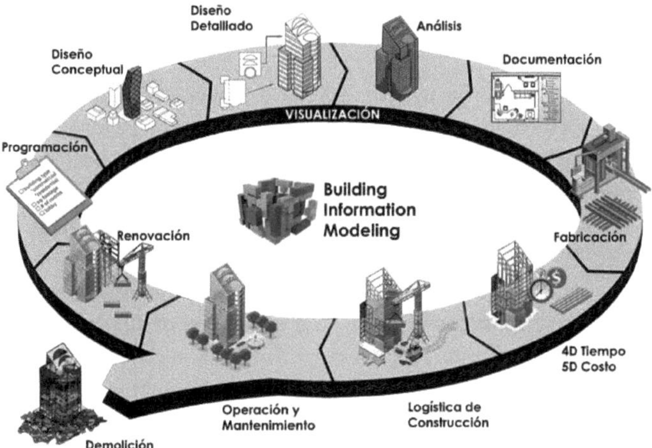

Source: BIM Space.

In short, BIM not only improves collaboration between disciplines, but also reduces errors and conflicts by detecting interferences, which in turn leads to time and cost savings throughout the project lifecycle. These advantages make BIM methodology an essential tool in the construction industry, promoting more efficient and effective project management.

4.4 Implementation process

The implementation of Building Information Modelling (BIM) in a project is a process that requires careful planning and methodical execution. This process involves not only the adoption of new technologies, but also a cultural change within the organisation. The following describes the steps to implement BIM, the training and education of personnel, as well as strategies to facilitate the transition from traditional methods to BIM.

4.4.1 Steps for Implementing BIM on a Project

BIM implementation starts with preparation and planning. The first step is to establish a BIM implementation plan that clearly defines the project objectives. This includes identifying specific goals to be achieved, such as improving efficiency, reducing costs or optimising collaboration between teams. It is essential that the company's management is involved in this phase, as their support is crucial for the success of the process (Eastman et al., 2011).

Once the objectives have been defined, the next step is to carry out a diagnosis of the current situation in the company. This involves assessing the existing processes, the tools used and the level of staff competence in relation to BIM. This diagnosis will help to identify areas that require improvement and to establish a timeline for implementation (Korman, 2010).

The next step is staff training. Training is essential to ensure that all team members understand the principles of BIM and how to apply them in their daily work. It is recommended to start with general training on the basics of BIM, followed by specific technical training on the tools to be used, such as Revit or Navisworks. This training should be continuous, as BIM is a constantly evolving field (Azhar, 2011).

Once the staff has been trained, the technical implementation must proceed. This includes the creation of an initial BIM model and the integration of the necessary data. It is advisable to start with a pilot project to test the methodology in a controlled environment before applying it to larger projects. This approach allows for adjustments and optimisations in the implementation process (Bynum et al., 2013).

Finally, it is crucial to establish a monitoring and evaluation system. This involves defining success indicators to measure the performance of BIM in the project. Continuous feedback and evaluation of the results will help to identify areas for improvement and adjust the approach as necessary (Sacks et al., 2010).

4.4.2 Staff Training and Education

Staff training is a critical component of the BIM implementation process. Without a well-trained team, even the best technology can be ineffective. Training should be comprehensive and tailored to the specific needs of each role within the team. For example, architects may need more in-depth training in design and modelling, while engineers may require training in systems analysis and coordination (Kiviniemi et al., 2012).

It is recommended that training be conducted in several phases. Firstly, basic training should be provided covering the fundamental principles of BIM and its importance in the construction industry. This initial training should include practical examples and case studies that demonstrate the benefits of BIM in real projects (Chong et al., 2017).

After basic training, technical training in the specific tools to be used on the project should be provided. This may include courses on software such as Revit, Navisworks and other modelling and information management software. In addition, it is important to encourage continuous training, as BIM is constantly evolving and new tools and techniques are developed regularly (Bynum et al., 2013).

Training should also include aspects of change management. This involves preparing staff to adapt to new ways of working and fostering a culture of collaboration and open communication. Resistance to change is a common challenge in implementing new technologies, so it is critical to address these concerns from the outset (Azhar, 2011).

4.4.3 Strategies for the Transition from Traditional Methods to BIM

The transition from traditional methods to BIM can be a significant challenge for many organisations. To facilitate this process, it is essential to develop strategies that address both the technical and cultural aspects of implementation. One of the most effective strategies is clear and consistent communication. Keeping all team members informed about the goals, benefits and progress of the implementation process will help reduce uncertainty and resistance to change (Sacks et al., 2010).

Another key strategy is the involvement of all levels of the organisation. From top management to operatives, everyone must be committed to the transition to BIM. This can be achieved through workshops, meetings and training sessions that include all departments involved in the construction process (Chong et al., 2017).

In addition, it is advisable to establish a BIM working group to take the lead in implementation and resolve issues as they arise. This group can be composed of representatives from different disciplines, ensuring that all perspectives are considered and collaboration is encouraged (Bynum et al., 2013).

Finally, it is important to recognise and celebrate achievements throughout the process. This not only motivates the team, but also reinforces the importance of BIM in the organisation. Celebrating milestones, such as the completion of a pilot project or the achievement of positive results, can help to consolidate the BIM culture within the company (Azhar, 2011).

In conclusion, BIM implementation is a complex process that requires careful planning, staff training and effective strategies for transitioning from traditional methods. By following these steps and fostering a culture of collaboration and continuous learning, organisations can take full advantage of the benefits BIM offers in construction project management.

4.5 BIM and sustainability

The integration of Building Information Modelling (BIM) methodology with sustainability principles has opened up new opportunities for resource optimisation and waste reduction in the construction sector. Through specific energy and environmental analysis tools, as well as examples of sustainable buildings designed with BIM, it can be seen how this methodology not only improves efficiency in design and construction, but also promotes more sustainable practices.

4.5.1 Tools for Energy and Environmental Analysis

The use of energy and environmental analysis tools is essential in the context of BIM, as it allows designers to evaluate the performance of a building prior to its construction. Software such as EnergyPlus, Green Building Studio and Ecotect are used to simulate and analyse the energy performance of buildings. These tools can be integrated with BIM models for real-time analysis, which facilitates the identification of areas for improvement in terms of energy efficiency and sustainability (Azhar, 2011).

For example, EnergyPlus is an advanced software that allows detailed simulations of a building's energy consumption, daylighting and thermal comfort. By integrating EnergyPlus with a BIM model, architects can adjust the design based on the results obtained, thus optimising the energy performance of the building (Bynum et al., 2013). In addition, tools such as Tally allow life cycle analysis (LCA) to be performed directly within a BIM model, helping designers to assess the environmental impact of design decisions from the earliest stages of the project.

4.5.2 Examples of Sustainable Buildings Designed with BIM

Several flagship projects have demonstrated the potential of BIM in the creation of sustainable buildings. One notable example is One Central Park in Sydney, Australia, which uses an innovative design that integrates vegetation into its structure. The use of BIM allowed architects and designers to optimise the building's orientation, maximise natural light and minimise energy consumption (Eastman et al., 2011). Thanks to the energy performance simulation, design solutions were achieved that improve building efficiency and promote a healthier living space.

Another example is the Bosco Verticale in Milan, Italy, a residential project that incorporates vertical gardens on its facades. The design was carried out using BIM to assess the sustainability of materials and the energy efficiency of the building. This approach not only contributes to the reduction of the carbon footprint, but also improves air quality and the well-being of the residents (Korman, 2010). These examples illustrate how BIM can be a powerful tool for designing buildings that not only meet sustainability standards, but also provide a healthier living environment.

4.5.3 BIM in Resource Optimisation and Waste Reduction

The impact of using BIM in optimising resources and reducing waste is significant. By enabling more accurate planning and improved coordination between disciplines, BIM helps minimise material waste during construction. The ability to detect interference and conflicts in the design phases allows teams to address problems before they become costly mistakes on the construction site (Sacks et al., 2010).

In addition, 3D modelling facilitates the visualisation and planning of material use, which contributes to more efficient resource management. For example, the ability to perform a life cycle analysis (LCA) allows designers to assess not only the initial cost of materials, but also their environmental impact throughout their lifetime. This encourages the selection of more sustainable materials and the implementation of construction practices that reduce waste (Azhar, 2011).

In terms of waste reduction, BIM models can facilitate the prefabrication of components, allowing for a more efficient use of materials and a reduction of waste generated at the construction site. By planning and manufacturing structural elements or systems in a controlled environment, material losses can be significantly reduced (Kiviniemi et al., 2012). This methodology not only improves project sustainability, but also optimises construction time and reduces associated costs.

4.6 Related Technological Innovations

The integration of Building Information Modelling (BIM) with emerging technologies such as artificial intelligence (AI) and augmented reality (AR) is transforming the construction industry. These innovations not only improve efficiency and accuracy in the design and execution of projects, but also enable better resource management and greater sustainability. In this context, it is essential to explore how these technologies are being implemented and the benefits they bring.

4.6.1 Integration of BIM with Artificial Intelligence

Artificial intelligence is becoming an essential component in the evolution of BIM. AI makes it possible to analyse large volumes of data generated during a project's lifecycle, making it easier to identify patterns and predict potential problems. For example, machine learning algorithms can be used to optimise building design, suggesting configurations that minimise energy consumption and maximise operational efficiency (Khan et al., 2020).

In addition, AI can automate repetitive tasks within the BIM process, such as element classification and collision detection. This not only reduces the time required to complete these tasks, but also decreases the likelihood of human error, resulting in a more accurate and efficient design (Zhou et al., 2019). AI's ability to perform predictive analytics also allows project teams to anticipate problems before they occur, which improves planning and risk management.

4.6.2 Integration of BIM with Augmented Reality

Augmented reality offers an innovative way to visualise construction projects in the context of the physical environment. By overlaying BIM models on the real world, professionals can interact with the design in a more intuitive way. This technology allows architects and contractors to see how a building will integrate into its environment before construction begins, making it easier to identify potential problems and make informed decisions (Dore & Murphy, 2017).

AR is also used to improve worker training at the construction site. Through immersive simulations, employees can familiarise themselves with procedures and equipment prior to actual construction, which reduces the risk of errors and improves safety on the job (Bae et al., 2020). This real-time visualisation capability not only improves communication between teams, but also enables more effective collaboration between architects, engineers and contractors.

4.6.3 Examples of Innovative Applications Combining BIM with IoT

The combination of BIM with the Internet of Things (IoT) is enabling the development of innovative solutions that improve building management and operational efficiency. The integration of IoT sensors into BIM models allows building managers to monitor the performance of systems in real time. For example, sensors can collect data on energy consumption, air quality and equipment status, enabling proactive management and identification of problems before they become costly failures (Gao et al., 2019).

One notable application of this integration is predictive maintenance. By analysing data collected by sensors, systems can predict when equipment needs maintenance, reducing downtime and improving operational efficiency. This not only optimises resource use, but also contributes to sustainability by minimising material and energy waste (Zhang et al., 2020).

In addition, the integration of IoT data into BIM models helps to optimise the use of resources during the construction and operation of buildings. For example, real-time data can be used to adjust a building's energy consumption, ensuring that only the necessary resources are used at any given time. This not only reduces operating costs, but also lowers the building's carbon footprint (Gao et al., 2019).

In conclusion, the integration of BIM with emerging technologies such as artificial intelligence, augmented reality and the Internet of Things is revolutionising the construction industry. These innovations not only improve efficiency and accuracy in the design and execution of projects, but also promote more sustainable practices and better resource management. As these technologies continue to evolve, we are likely to see an even greater impact on the way buildings are designed, constructed and managed.

4.7 Challenges and limitations of BIM

The adoption of Building Information Modelling (BIM) has transformed the construction industry, offering significant benefits in terms of efficiency, collaboration and sustainability. However, its implementation is not without its challenges and limitations. This article explores common barriers to BIM adoption, such as cost and resistance to change, and proposes solutions and strategies to overcome these obstacles.

4.7.1 Common Barriers to BIM Adoption

High Initial Cost: One of the main barriers to BIM adoption is the initial cost associated with implementing this technology. This includes investment in software, hardware and staff training. Many firms, especially small and medium-sized ones, may find it difficult to justify this initial investment, which limits their ability to adopt BIM (Azhar, 2011).

Resistance to Change: Resistance to change is another significant barrier. Many construction professionals are accustomed to traditional methods and may be reluctant to adopt new technologies. This resistance may manifest itself in a lack of interest in training or in implementing new processes (Sacks et al., 2010).

Lack of Training and Knowledge: Lack of adequate BIM skills and knowledge is a common challenge. Successful implementation of BIM requires employees to be trained in the use of specific software and modelling principles. Without adequate training, firms may struggle to take full advantage of BIM capabilities (Bynum et al., 2013).

Interoperability and Standards: Interoperability between different BIM platforms and software is a major challenge. Lack of consistent standards can hinder collaboration between different disciplines and companies, which limits the effectiveness of BIM in complex projects (Kiviniemi et al., 2012).

Data Security Concerns: Data security is a growing concern in the adoption of BIM. Firms may be reluctant to share sensitive information in a digital environment, which can hinder collaboration and transparency in projects (Eastman et al., 2011).

4.7.2 Solutions and Strategies to Overcome Challenges

Long-Term Cost Justification: To address the initial cost challenge, firms must focus on long-term return on investment (ROI) justification. This involves demonstrating how BIM implementation can reduce operational costs, minimise errors and improve efficiency over the project lifecycle. Case studies showing the success of BIM in previous projects can be useful to convince stakeholders (Bynum et al., 2013).

Change Management: Change management is crucial to overcome resistance to change. Firms should implement change management strategies that include clearly communicating the benefits of BIM, involving employees in the adoption process and creating a supportive environment. Ongoing training and professional development are also essential to facilitate the transition (Sacks et al., 2010).

Training and Development Programmes: Investing in training and development programmes is critical to addressing the BIM skills gap. Firms should provide practical and accessible training for their employees, ensuring that everyone has the opportunity to acquire the skills needed to use BIM effectively. This can include workshops, online courses and certifications (Kiviniemi et al., 2012).

Establishing Standards and Protocols: To improve interoperability, it is important that companies and industry organisations work together to establish common standards and protocols for the use of BIM. This will facilitate collaboration between different disciplines and companies, improving the efficiency and effectiveness of projects (Eastman et al., 2011).

Data Security and Trust: To address data security concerns, companies must implement robust and transparent security measures. This includes the use of encryption technologies, training staff on security practices and creating clear policies on handling sensitive data. Fostering a culture of trust and transparency is also essential to facilitate collaboration (Azhar, 2011).

Thus, the adoption of BIM in the construction industry presents significant challenges, but also offers opportunities to improve efficiency and sustainability. By addressing common barriers, such as cost, resistance to change and lack of training, firms can maximise the benefits of BIM and remain competitive in an ever-changing marketplace. Implementing effective change management, training and standard setting strategies is critical to overcoming these challenges and ensuring successful BIM adoption.

4.8 Future of BIM

Building Information Modelling (BIM) has revolutionised the construction industry, providing a more efficient and collaborative approach to the design, construction and management of buildings. As technology advances, BIM is also evolving, incorporating new dimensions and capabilities that extend its usefulness across the building lifecycle. This article explores emerging trends in the construction industry related to BIM, the evolution towards BIM 4D and BIM 5D, and the potential of BIM in the management of buildings throughout their lifecycle.

4.8.1 Emerging Trends in Construction Related to BIM

One of the most prominent trends is the adoption of BIM in the cloud. This migration to cloud-based platforms allows easier and more collaborative access to models, facilitating real-time updates and improving communication between all project stakeholders. This accessibility not only improves efficiency, but also reduces the risk of errors due to outdated information (Zhang et al., 2021).

Another significant trend is the use of digital twins. These are virtual replicas of physical assets that allow professionals to simulate and analyse the behaviour of a building in different scenarios. Digital twins provide real-time data that can optimise building performance and forecast maintenance needs, resulting in more effective resource management (Krygiel & Nies, 2016).

The integration of the Internet of Things (IoT) is also changing the way buildings are managed. With the addition of IoT sensors, it is possible to collect real-time data on the performance of systems such as heating, ventilation and air conditioning (HVAC) and security. This data is integrated with BIM models to provide a more complete view of building performance, improving decision-making and operational efficiency (Gao et al., 2019).

Artificial intelligence (AI) is beginning to play a crucial role in process optimisation within BIM. Through the analysis of large volumes of data, AI can identify patterns and anticipate problems before they occur, allowing construction teams to make more informed and proactive decisions (Bynum et al., 2013). In addition, automation in construction, which includes the use of robots for specific tasks, is gaining ground. This trend not only improves efficiency, but also reduces safety risks by minimising human intervention in dangerous tasks (Sacks et al., 2010).

4.8.2 Evolution of BIM towards 4D BIM and 5D BIM

The evolution of BIM has led to the creation of new dimensions that add significant value to project management. 4D BIM incorporates the dimension of time, allowing construction teams to plan and visualise the project schedule more effectively. This integration allows the construction process to be simulated, helping to identify potential conflicts and optimise the sequence of activities. By visualising the schedule in a BIM environment, all stakeholders can better understand the progress of the project, improving communication and collaboration between teams (Eastman et al., 2011).

On the other hand, 5D BIM adds the dimension of cost to the models, enabling more accurate financial management of the project. With BIM 5D, design changes are automatically reflected in cost estimates, allowing teams to make informed budget decisions at any time. This integration facilitates budget tracking throughout the project lifecycle, allowing teams to identify deviations and adjust the approach as needed (Zhang et al., 2021).

4.8.3 Potential of BIM in Building Management

BIM is not only useful during the design and construction phase, but also has significant potential in managing buildings throughout their lifecycle. By integrating IoT sensor data with BIM models, facility managers can anticipate maintenance needs before they become costly problems. This not only improves operational efficiency, but also extends the useful life of assets (Gao et al., 2019).

In addition, BIM enables more effective space management within a building. Models can be used to analyse space utilisation and make adjustments to optimise functionality and efficiency. It also facilitates the management of documentation needed to comply with standards and regulations, generating reports more efficiently, which reduces the administrative burden (Kiviniemi et al., 2012).

Sustainability and energy efficiency are other areas where BIM can have a significant impact. BIM models can simulate a building's energy performance, allowing managers to identify opportunities to improve energy efficiency and reduce operating costs. As buildings age, BIM can be used to plan renovations and upgrades, helping managers make informed decisions about future investments (Bynum et al., 2013).

So, the future of BIM is promising, with trends continuing to transform the construction industry. The evolution towards 4D BIM and 5D BIM offers new opportunities to improve project planning and management, while the potential of BIM in lifecycle building management promises to optimise efficiency and sustainability. It is essential that building professionals embrace these innovations to remain competitive and maximise the value of their projects.

4.9 Case studies

4.9.1 Hong Kong International Airport (HKIA)

One of the most emblematic cases is Hong Kong International Airport (HKIA). During its expansion, the implementation of BIM was crucial to manage the complexity of the project. Collaboration between different disciplines, such as architecture, engineering and construction, was significantly improved through BIM integration. This allowed for greater consistency in the design and execution of the project. In addition, the use of 3D visualisation and simulations helped to identify and resolve problems before the construction phase, resulting in a significant reduction of errors and rework. It was estimated that the use of BIM at the HKIA resulted in a 10% saving in construction costs and a 15% reduction in project delivery time, allowing the airport to begin operations ahead of schedule (Eastman et al., 2011).

4.9.2 One World Trade Center Building, New York

This project not only had symbolic significance, but also involved considerable technical challenges. During its construction, BIM was used to coordinate the multiple disciplines involved, allowing safety risks to be effectively managed. By modelling the site and construction activities, the team was able to foresee dangerous situations and plan preventive actions. In addition, BIM facilitated better planning of resource use, resulting in reduced material waste and improved operational efficiency. Communication also benefited from this technology, as visualisation and collaborative modelling improved coordination between contractors and subcontractors. This resulted in smoother project execution, reducing delays and conflicts.

4.9.3 The University of Health Sciences Hospital, London

In healthcare, the University of Health Sciences Hospital in London is an example of how BIM can transform the construction of medical facilities. The implementation of BIM was instrumental in addressing the design and functionality challenges in such a critical environment. Thanks to detailed simulations of patient and staff flow, the architects and engineers were able to optimise the hospital design, improving both functionality and user experience. In addition, an information model was created that covered not only the construction, but also the long-term maintenance of the hospital's equipment and systems. This facilitated more proactive and efficient facility management, allowing the hospital to reduce operating costs by 20% through better resource management and reduced equipment downtime (Bynum et al., 2013).

4.9.4 The Crossrail Project, London

With a total cost in excess of £20 billion, the implementation of BIM has been instrumental in coordinating the multiple phases of the project. The extension of the rail network includes numerous stations and tunnels, making construction management particularly complex. The use of BIM enabled teams to coordinate construction activities across multiple contractors, minimising conflicts and delays. Through 4D simulation, teams were able to plan and visualise the construction sequence, resulting in more efficient execution and fewer disruptions to existing operations. In addition, the use of BIM models provided stakeholders with a clear view of the project's progress, increasing transparency and trust between the parties involved.

The benefits of implementing BIM on these projects are clear. In the case of Hong Kong International Airport, cost and time reduction not only allowed the project to be completed ahead of schedule, but also improved the user experience. At One World Trade Center, security management and resource optimisation not only contributed to a safer working environment, but also resulted in a more efficient use of materials. Meanwhile, the University of Health Sciences Hospital showed how BIM can facilitate not only construction, but also the long-term management of the facility, directly impacting on the reduction of operating costs.

Crossrail, meanwhile, is an example of how collaboration between multiple disciplines and contractors can be optimised through the use of BIM. The ability to visualise and simulate processes in a digital environment allowed teams to anticipate and resolve conflicts before they became real problems on the construction site.

In summary, examples of projects such as Hong Kong International Airport, One World Trade Center, University of Health Sciences Hospital and Crossrail demonstrate how the adoption of BIM can transform the construction industry. The implementation of this technology not only improves efficiency and collaboration between teams, but also provides significant benefits in terms of cost reduction, time management and resource optimisation. The ability to anticipate problems, optimise processes and improve communication is crucial to the success of complex projects in an industry that faces constant challenges. With the continued advancement of technology and the increasing adoption of BIM, we are likely to see even more innovations and improvements in the construction industry in the future.

Bibliographical References

Al-Tabbaa, A., & Azzam, R. (2015). *Self-repairing concrete: A review of the state of the art. Construction and building materials, 98,* 1-10. https://doi.org/10.1016/j.conbuildmat.2015.05.001

SINC Agency (2023). *Carbon nanotubes to monitor public works from the inside.* https://www.agenciasinc.es/Noticias/Nanotubos-de-carbono-para-monitorizar-desde-dentro-las-obras-publicas

Azhar, S. (2011). *Building information modeling (BIM): Trends, benefits, risks, and challenges for the architecture, engineering and urban planning sector. Engineering Leadership and Management,* 11(3), 241-252.

Bae, J., Lee, J., & Kim, J. (2020). *Augmented reality for construction safety: A review. Automation in Construction*, 113, 103139. https://doi.org/10.1016/j.autcon.2020.103139

Böngen, A., et al. (2018). *Self-healing concrete: A review of the state of the art. Construction and Building Materials, 174,* 1-12. https://doi.org/10.1016/j.conbuildmat.2018.04.042

Bynum, P., Issa, R. R. A., & Karim, A. (2013). *Building information modeling in support of sustainable design and construction. Journal of Construction Engineering and Management*, 139(1), 1-8.

Argentine Chamber of Construction (2019). *Nanotechnology in the construction industry.*https://biblioteca.camarco.org.ar/PDFS/serie%2037/L9-%20IyT_NANOTECNOLOGIA%20EN%20LA%20INDUSTRIA%20DE%20%20LA%20CONTRUCCION%20(LIBRO.CD).pdf

Chong, H. Y., Lee, Y. S., & Wang, X. (2017). The role of building information modeling in enhancing collaboration in construction projects: A literature review. International Journal of Project Management, 35(3), 401-411.

CIMAV. (2023). *Area of Nanostructures and Polymeric Nanocomposites* https://cimav.edu.mx/investigacion/subsede-monterrey/area-de-nanoestructuras-y-nanocompuestos-polimericos/

Building Better Projects (2018, October 2). *Nanotechnology in precast concrete.* https://construyendomejoresproyectos.blogspot.com/2018/10/nanotecnologia-en-los-concretos.html

Dooko (2023). *Carbon Nanotubes: The technology is coming to the sector.* https://dooko.es/nanotubos-de-carbono-dooko/

Dore, C., & Murphy, M. (2017). *The role of augmented reality in the construction industry. Journal of Information Technology in Construction*, 22, 1-16. https://www.itcon.org/2017/1

Eastman, C., Teicholz, P., Sacks, R., & Liston, K. (2011). *BIM Handbook: A Guide to Building Information Modeling for Owners, Managers, Designers, Engineers, and Contractors. Wiley.*

Expocihachub (2023). *Nanoparticles can turn cement into an electrical conductor.* https://www.expocihachub.com/nota/tecnologia/nanoparticulas-pueden-convertir-cemento-en-conductor-electrico

Gao, S., Zhang, Y., & Wang, Y. (2019*). The integration of BIM and IoT for smart construction: A review. Journal of Building Performance*, 10(3), 1-12. https://doi.org/10.21834/jbp.v10i3.327

Grinder (2023). *Nanotechnology in the construction industry.* https://grinder.cl/nanotecnologia-en-la-industria-de-la-construccion/

Henkensiefken, J. R., & Schlangen, E. (2015). *The potential of self-healing concrete for sustainable construction.* Journal of Cleaner Production, *112,* 1-10. https://doi.org/10.1016/j.jclepro.2015.01.001

Hormigón al Día. (2023). *How to improve the strength of cement? With nanoparticles from shrimp shells.* https://hormigonaldia.ich.cl/smartconcrete/como-mejorar-la-resistencia-del-cemento-con-nanoparticulas-del-caparazon-de-camarones/

Huang, Y., et al. (2015). Self-healing concrete: A review of the state of the art. *Construction and Building Materials, 93,* 1-12. https://doi.org/10.1016/j.conbuildmat.2015.05.028

Imnovation Hub (2023). *Self-cleaning paint as a response to air pollution.* https://www.imnovation-hub.com/es/construccion/pintura-autolimpiable-purificante/

İpek, S., Güneyisi, E. M., & Güneyisi, E. (2023). *Bacteria-based self-healing concretes for sustainable* structures. https://doi.org/10.1201/9781003325246-10

National Institute for Occupational Safety and Health (2023). *Nanomateriales.* https://www.insst.es/documents/94886/96076/sst+nanomateriales/bd21b71f-d5ec-4ee8-8129-a4fa58480968

Jonkers, H. M. (2011). Bacteria-based self-healing concrete. *Heron, 56*(1), 1-12.

Kaur, T., & Singh, A. (2023). *A review on self healing concrete.* International Journal for Research in Applied Science and Engineering Technology, 11. https://doi.org/10.22214/ijraset.2023.54736

Krygiel, E., & Nies, B. (2016). *BIM Handbook: A Guide to Building Information Modeling for Owners, Managers, Designers, Engineers, and Contractors. Wiley.*

Kakooei, S., et al. (2017). Sustainability of self-healing concrete: A review. *Journal of Cleaner Production, 142,* 1-12. https://doi.org/10.1016/j.jclepro.2016.11.086

Khan, M. A., Ali, A., & Khan, M. A. (2020). *Artificial intelligence in construction: A review. Journal of Civil Engineering and Management*, 26(5), 453-467. https://doi.org/10.3846/jcem.2020.12492

Kiviniemi, M., Fischer, M., & Hartmann, T. (2012). *Building information modeling for sustainable design. Journal of Green Building*, 7(2), 50-66.

Korman, T. (2010). *BIM for Facility Managers*. Wiley.

Le, T. M., et al. (2012). Self-healing concrete with polymeric materials. *Materials and Structures, 45*(1), 1-12. https://doi.org/10.1617/s11527-011-9791-4

Lesovik, V., Fediuk, R., Amran, M., Vatin, N., & Timokhin, R. (2021). Self-healing construction materials: *The geomimetic approach. Sustainability, 13,* 9033. https://doi.org/10.3390/su13069033

Le, T. M., et al. (2012). Self-healing concrete with polymeric materials. *Materials and Structures, 45*(1), 1-12. https://doi.org/10.1617/s11527-011-9791-4

Lepech, M. D., & Li, V. C. (2008). Autonomous healing of concrete: A review of the state-of-the-art. *Journal of Materials in Civil Engineering, 20*(9), 1-10. https://doi.org/10.1061/(ASCE)0899-1561(2008)20:9(1)

Marcondes, F. C., et al. (2012). Carbon nanotubes in Portland cement concrete. *Influence of addition on mechanical properties*. Revista Mexicana de Ingeniería Química, 10(2), 97-104.

Mechtcherine, V., et al. (2016). Self-healing concrete: A review of the state of the art. *Materials and Structures, 49*(1), 1-12. https://doi.org/10.1617/s11527-015-0660-2

Nanobiology (2023). *Nanotechnology and its impact on the environment.* https://nanobiologia.com/blog/nanotecnologia-y-su-impacto-en-el-medio-ambiente

Ramakrishnan, V., et al. (2009). Self-healing concrete: A new approach to sustainable construction. *Journal of Materials in Civil Engineering, 21*(7), 1-8. https://doi.org/10.1061/(ASCE)0899-1561(2009)21:7(1)

Ravikar, A., Joshi, D., & Menon, R. (2023). Analysis of self-healing strategies in smart concrete using fuzzy analytic hierarchy process. *E3S Web of Conferences, 405.* https://doi.org/10.1051/e3sconf/202340504016

Repsol (2023). *Nanotechnology: What it is, Applications and its 4 types.* https://www.repsol.com/es/energia-futuro/tecnologia-innovacion/nanotecnologia/index.cshtml

Redalyc. (2023). *Impact-resistant polymer-based nanocomposites.* https://www.redalyc.org/journal/4435/443562640002/html/

Redalyc (2015). *Nanotechnology and the environment: environmental implications.* https://www.redalyc.org/pdf/1794/179420814007.pdf

Rickerby et al. (2023). *Nanotechnology for the environment: beauty not the beast.* https://cordis.europa.eu/article/id/27711-nanotechnology-and-the-environment-beauty-rather-than-beast/es

Sacks, R., Eastman, C., Lee, G., & Jeong, Y. (2010). The role of BIM in the construction industry. Automated Construction, 19(3), 201-210.

SciELO (2017). *Synthesis of polymer nanocomposites with graphene and their properties.* http://www.scielo.org.pe/scielo.php?pid=S1810-634X2017000100007&script=sci_arttext

StudySmarter ES (2023). *Self-Cleaning Materials: Technology, Nanotechnology* - StudySmarter EN. https://www.studysmarter.es/resumenes/estudios-de-arquitectura/construccion/materiales-autolimpiantes/

Tittelboom, K., & De Belie, N. (2013). Self-healing in cementitious materials-A review. *Materials, 6,* 2182-2217. https://doi.org/10.3390/ma6062182

The Royal Society (2004). *Nanoscience and nanotechnologies: opportunities and uncertainties.* https://royalsociety.org/-/media/Royal_Society_Content/policy/publications/2004/9246.pdf

Tzortzopoulos, P., Kagioglou, M., & Cooper, R. (2011). The role of building information modeling in the construction process: A case study of a large-scale project. Construction Innovation, 11(3), 287-305.

UCL Chemistry (2023). *Nanotechnology and architecture: self-cleaning paints.* https://arquitecturayempresa.es/noticia/nanotecnologia-y-arquitectura-pinturas-autolimpiables

Polytechnic University of Madrid (2017). *Nanotechnology in architecture: Graphene.* https://oa.upm.es/50243/1/INVE_MEM_2017_272133.pdf

European University (2023). *Functionalised carbon nanotubes improve the mechanical properties of construction materials such as cement.* https://universidadeuropea.com/noticias/los-nanotubos-de-carbonofuncionalizados-mejoran-las-propiedades-mecanicas-de-los-materiales-de-cons/

Van Tittelboom, K., et al. (2010). Self-healing concrete: A review of the state of the art. *Construction and Building Materials, 24*(3), 1-12. https://doi.org/10.1016/j.conbuildmat.2009.09.00

Ve et al. (2015). *Nanoparticles and the environment: key considerations.* https://ve.scielo.org/scielo.php?pid=S1316-48212015000100005&script=sci_arttext

Van Tittelboom, K., et al. (2010). Self-healing concrete: A review of the state of the art. *Construction and Building Materials, 24*(3), 1-12. https://doi.org/10.1016/j.conbuildmat.2009.09.007

Wang, J., et al. (2016). Self-healing concrete with microcapsules: A review. *Construction and Building Materials, 112,* 1-12. https://doi.org/10.1016/j.conbuildmat.2016.02.045. https://doi.org/10.1016/j.conbuildmat.2016.02.045

Yao Lu et al. (2023). *Self-cleaning surfaces: towards the future of cleaning and disinfection.* https://higieneambiental.com/aire-agua-y-legionella/superficies-autolimpiables-hacia-el-futuro-de-la-limpieza-y-desinfeccion

Zavala Murguía, J. (2023). *Study on polymer nanocomposites: properties and potential applications.* CIQA Repository. https://ciqa.repositorioinstitucional.mx/jspui/bitstream/1025/379/1/Juan%20Zavala%20Murguia.pdf

Zhang, Y., et al. (2016). Nanomaterials for self-healing concrete: A review. *Materials Science and Engineering: A, 674,* 1-12. https://doi.org/10.1016/j.msea.2016.05.045

Zhao, K., Ma, X., Zhang, H., & Dong, Z. J. (2022). Performance zoning method of asphalt pavement in cold regions based on climate indexes: A case study of Inner Mongolia, China. *Construction and Building Materials, 361,* 129650. https://doi.org/10.1016/j.conbuildmat.2022.129650

Zhou, Y., Wang, Y., & Zhang, Y. (2019). The application of artificial intelligence in construction: A review. Journal of Construction Engineering and Management, 145(5), 04019020. https://doi.org/10.1061/(ASCE)CO.1943-7862.0001670.

Zhang, Y., Gao, S., & Wang, Y. (2020). IoT-based smart construction: A review. Journal of Building Performance, 11(1), 1-12. https://doi.org/10.21834/jbp.v11i1.327

Printed by Books on Demand GmbH, Norderstedt / Germany